Banach Spaces of Analytic Functions

CONTEMPORARY MATHEMATICS

454

Banach Spaces of Analytic Functions

AMS Special Session
April 22–23, 2006
University of New Hampshire
Durham, New Hampshire

Rita A. Hibschweiler
Thomas H. MacGregor
Editors

American Mathematical Society
Providence, Rhode Island

2000 *Mathematics Subject Classification.* Primary 30D15, 30D45, 30D50, 30D55, 30E20, 30H05, 32A18, 46E15, 47B33.

Library of Congress Cataloging-in-Publication Data

Banach spaces of analytic functions : AMS Special Session, April 22–23, 2006, University of New Hampshire, Durham, New Hampshire / Rita A. Hibschweiler, Thomas H. MacGregor, editors.
p. cm. — (Contemporary mathematics, ISSN 0271-4132 ; v. 454)
Includes bibliographical references.
ISBN 978-0-8218-4268-3 (alk. paper)
1. Banach spaces—Congresses. 2. Analytic functions—Congresses. I. Hibschweiler, Rita A. II. MacGregor, T. H. (Thomas H.). III. Title.

QA322.2.B315 2008
515′.732—dc22 2007060849

---

10 9 8 7 6 5 4 3 2 1 13 12 11 10 09 08

This volume is dedicated to the memory of Alec Matheson. Alec was a major contributor to this field, and he collaborated with several of the authors represented in this volume. Alec is remembered as a highly respected mathematician, valued colleague and friend.

Contents

Preface

Banach spaces of analytic functions play an important role in both classical and modern analysis. In this Proceedings, the major focus is on spaces of functions analytic in the open unit disc, such as the Hardy spaces, the Bergman spaces, and the weighted versions of these spaces. Other spaces under consideration here include the Bloch space, the families of Cauchy transforms and fractional Cauchy transforms, BMO, VMO, and the Fock space. Some of the work deals with questions in several complex variables.

One of the themes of the Proceedings is the notion of multiplication operators, composition operators and weighted composition operators acting on Banach spaces. This has been a topic of extensive research over the past twenty years. The Proceedings includes results characterizing bounded, compact and isometric composition operators in various settings.

The recently published book "The Cauchy Transform" (2006) by Joseph A. Cima, Alec L. Matheson and William T. Ross, deals with material that is central to this Proceedings. A related book "Fractional Cauchy Transforms" (2006) was written by the editors of this Proceedings. Research by these authors is part of the Proceedings. The articles here were generated by the Special Session on Banach Spaces of Analytic Functions, held as part of the regional AMS meeting at the University of New Hampshire in April, 2006. It was very unfortunate that Alec Matheson, a major contributor to this field, died shortly before the meeting after a long illness. A brief note on the life of Alec Matheson, including several of his results with various collaborators, was prepared by Joseph Cima and is included here.

This volume presents both expository and research articles. Several authors conclude their articles with open questions.

The Special Session on Banach Spaces of Analytic Functions brought together thirty mathematicians, mostly from the northeastern United States. We hope that this volume serves to communicate more broadly the research presented at the meeting.

Contemporary Mathematics
Volume **454**, 2008

A NOTE ON THE LIFE OF ALEC MATHESON AND SOME OF HIS WORK

JOSEPH A.CIMA

This short note concerns parts of the personal and professional life of my friend and frequent co-author Alec Matheson. I have included the statements of some of the work that Alec accomplished in his professional life.

Alec was born near Aberdeen, Washington on October 26, 1946. Alec's father had been in World War II and during that period his mother worked in a factory helping to produce ships. After the war his father was a glazier and ran his own shop. Alec worked for his father in his teenage years. Alec was very attached to his parents and siblings. He returned to Aberdeen to visit his parents almost every summer and at Christmas time.

Alec attended the University of Washington in Seattle and graduated with his B.S. degree in 1968. He developed a love for languages, history and politics. He pursued all of these in addition to his mathematical career. It was not unusual for him to bring up a subject like "the Third Years War" for discussion. He read history avidly.

After leaving Seattle he decided to attend the University of Illinois at Urbana. During this time his work towards the Doctorate was interrupted by service in Vietnam. This was one of the deepest and affecting parts of Alec's life. He was awarded the Bronze Star, Vietnam Campaign Medal and a Vietnam Service Medal. This service time and his experiences in Vietnam affected him seriously for the remainder of his life. In one of our working visits to a coffee shop in Chapel Hill, he became quite disturbed and eventually came under a doctor's care for many weeks.

Alec finished his degree under the supervision of Robert Kaufman. The acknowledgement on his Ph.D. thesis reads as follows: "I would like to thank Professor Robert Kaufman for his assistance, encouragement and patience in the preparation of this thesis, and Professor J. (Jerry) Uhl, Jr. for his belief that it could be done." He carried this admiration and high regard for his teachers during his entire professional life. The title of his thesis is "Closed ideals in Banach algebras of analytic functions satisfying a Lipschitz condition." A study of the references for his thesis shows five of these are publications of authors from the U.S.S.R., and four of these are from the Soviet presses. His skill with this language was a positive, important attribute for this work and many to come.

I had news of his thesis results and went to a conference at Kent State to hear him talk about them. One of the results coming out of this work is the following [1]. For $0 < \alpha < 1$ let λ_α denote the class of functions f analytic in the unit disk, continuous in the closed disk for which $t^{-\alpha}\omega(t) \to 0$ as $t \to 0$, where ω denotes the modulus of continuity of the boundary function of f $\left(\text{e.g., } \omega(t) \equiv \sup_{|\zeta-\zeta'|\le t} |f(\zeta) - f(\zeta')|\right)$. Defining a norm

---

Date: September 25, 2007.
1991 *Mathematics Subject Classification.* 01A70.

as

$$\|f\|_{\alpha} = \|f\|_{\infty} + \sup t^{-\alpha}\omega(t),$$

the space λ_α becomes a Banach algebra.

Theorem A. *Let $f \in \lambda_\alpha$ and let E be a closed set on the unit circle such that $f(z) = 0$ for all $z \in E$. Let $M > 0$ be given. Then for every $\varepsilon > 0$ there exists a function $f_\alpha \in \lambda_\alpha$ such that*

(i) the inner factors of f and f_α coincide,

(ii) $\|f - f_\alpha\| < \varepsilon$, and

(iii) $|f_\alpha(z)| = 0\,(dist^M(z, E))$ as dist $(z, E) \to 0$.

This theorem can be used to give a characterization of the closed ideals in λ_α analogous to the Rudin-Beurling characterization of the closed ideals in the disc algebra.

Alec's results interested me a great deal and I asked him to visit the University of North Carolina. He spent a semester at Chapel Hill and that is when our first collaboration took place. We proved in [2] the following.

Theorem B. *Let Q be a crescent bounded by two internally tangent circles and let ψ be a conformal mapping of the unit disk D onto Q. Then ψ satisfies*

$$\sup\left\{\frac{1}{|S|}\int_S |\psi'|^{2-p}\right\}^{1/p}\left\{\frac{1}{|S|}\int_S |\psi'|^{2-q}\right\}^{1/q} < \infty,$$

for any Carleson rectangle S contained in the unit disc. The numbers p and q are conjugate Hölder indices.

This result can be used to prove that for $1 < p < \infty$ the Bergman projection P_Q is bounded on $L^P(Q)$.

Alec spent several years at Oklahoma State University. While there he collaborated with Dale Alspach and Joe Rosenblatt on projections to translation-invariant subspaces of $L_1(G)$ [3]. He also worked with David D. Ullrich on weighted averages in Hardy spaces.

After leaving Oklahoma State University he went to Lamar University. Although Lamar did not offer as many research opportunities as Oklahoma State, he kept up his research interests. In particular he worked with John Cannon on free boundary value problems related to the combustion of a solid [4]. Altogether Alec published 37 papers and one book. In addition to those mentioned above he has worked with his colleague Valentin Andreev as well as Kevin Madigan, Alexander Pruss, Paul Bourdon and Bill Ross.

Alec took advantage of an offer to visit SUNY at Albany. This was arranged by Michael Stessin. He really enjoyed his year at SUNY and Michael and he finished two papers together (see, for example, [5]).

Alec spent many weeks visiting me in my home both at Chapel Hill and in Missoula, Montana. Part of this time together was work but we both enjoyed walking in the Rockies and the Smokies.

Finally, Alec, Bill Ross and I worked on material related to Cauchy transforms. In addition to two papers on this subject Alec, Bill and I worked diligently on a recent book *The Cauchy Transform* [6].

Let me add a few more theorems that occurred during the period when Alec and I were working closely together. One that he was really happy with occurs in the paper [7]. It relates to composition operators C_ϕ on Hardy spaces and the work by D. Sarason and work by Joel Shapiro and C. Sundberg on compactness of such operators. We were able to give a "direct function theoretic" proof of some of this material.

One of the last areas we planned to consider was a study of weak type inequalities for functions from the classical Banach spaces of analytic functions. In particular we revisited a result of D. Bekollé which stated that the Bergman projection $\mathbf{P}$ on functions in $L^1(\mathbb{D}, dA)$ satisfied a weak type estimate of the form

$$\text{area}(z \in \mathbb{D} : |\mathbf{P}f(z)| > t) \leq \frac{C}{t}\|f\|_1.$$

Bekollé's result is valid for the Bergman space on the ball in $\mathbb{C}^m$ as well. We have shown the following (unpublished).

Theorem C. *Let μ be a finite Borel measure on $\mathbb{D}$ and let $\mathbf{P}\mu$ denote the Bergman projection of μ, that is,*

$$\mathbf{P}\mu(z) = \int_{\mathbb{D}} \frac{d\mu(\omega)}{(1 - z\bar{\omega})^2}.$$

Then $\mathbf{P}\mu$ satisfies the weak type estimate

$$\textit{area}(z \in \mathbb{D} : |\mathbf{P}\mu(z)| > t) \leq \frac{C}{t}\|\mu\|,$$

where C is independent of μ. This is a consequence of a more general result. Let Ω be a domain in the complex plane and let

$$K : \Omega \times \Omega \to \mathbb{C}$$

be a function such that for each fixed $z \in \Omega$, the function $K(z, \cdot)$ is bounded and harmonic in Ω. Then for every finite Borel measure μ on Ω, the integral

$$f_\mu(z) = \int_\Omega K(z, \zeta)\, d\mu(\zeta)$$

satisfies a weak type 1:1 inequality $\Big($ e.g., area $\{z \in \Omega \| f_\mu(z)| > t\} \leq \frac{c}{t}\|\mu\|\Big)$.

Lemma. *For every finite Borel measure μ on Ω, there is a function $f \in L^1(\Omega, dA)$ such that $\|f\|_1 = \|\mu\|$ and*

$$f_\mu(z) = \int_\Omega K(z, \zeta)\, f(\zeta)\, dA(\zeta)$$

for every $z \in \Omega$.

Let me end this note with an apology to any of Alec's co-authors and friends whose works I have slighted in this note. Those of us who worked closely with Alec (especially Bill Ross and Michael Stessin) can attest to Alec's ability in mathematics, his hard work, care and deep commitment to his research and his students. During these latter years when Alec had his operations for cancer and was taking heavy chemo-therapy he never complained. He worked like a trooper until the end. His close friend Valentin (Andreev) found him on the floor of his home and rushed him to hospital where he spent his last days. Ross and I spoke to him by phone the last ten days of his life. Again and again he would give me a short rundown on his condition and then tell me about a mathematical result we should think about. He received the printed copy of our joint book on the morning he passed away.

He is missed for his ability and his humanity.

REFERENCES

[1] Matheson, A. Approximation of analytic functions satisfying a Lipschitz condition. *Michigan Math. J.* **25** (1978), no. 3, 289–298.

[2] Cima, J. A.; Matheson, A. Approximation in the mean by polynomials. *Rocky Mountain J. Math.* **15** (1985), no. 3, 729–738.

[3] Alspach, D.; Matheson, A.; Rosenblatt, J. Projections onto translation-invariant subspaces of $L_1(G)$. *J. Funct. Anal.* **59** (1984), no. 2, 254–292.

[4] Cannon, John R.; Matheson, Alec L. A free boundary value problem related to the combustion of a solid: flux boundary conditions. *Quart. Appl. Math.* **55** (1997), no. 4, 687–705.

[5] Matheson, Alec L.; Stessin, Michael I. Cauchy transforms of characteristic functions and algebras generated by inner functions. *Proc. Amer. Math. Soc.* **133** (2005), no. 11, 3361–3370 (electronic)

[6] Cima, Joseph A.; Matheson, Alec L.; Ross, William T. *The Cauchy transform.* Mathematical Surveys and Monographs, 125. American Mathematical Society, Providence, RI, 2006.

[7] Cima, Joseph A.; Matheson, Alec L. Essential norms of composition operators and Aleksandrov measures. *Pacific J. Math.* **179** (1997), no. 1, 59–64.

UNC, CHAPEL HILL, N.C. 27599

Contemporary Mathematics
Volume **454**, 2008

ON THE INVERSE OF AN ANALYTIC MAPPING

JOSEPH A.CIMA

Section 1.

Assume Ω is a domain in a separable Hilbert space $\mathbb{H}$ and f is an analytic mapping from $\Omega \subset \mathbb{H}$ into $\mathbb{H}$. This statement means that for each $a \in \Omega$, there is associated (in a continuous way) to f a bounded linear operator (the Frechet derivative) $Df(a)$ satisfying

$$\lim_{b \to a} \frac{\|f(b) - f(a) - Df(a)(b-a)\|}{\|a-b\|} = 0.$$

If $\mathbb{H}$ is finite dimensional, say $\dim \mathbb{H} = n$, we may assume $\mathbb{H} = \mathbb{C}^n$, and that f is an analytic mapping of a domain $\Omega \subset \mathbb{C}^n \to \mathbb{C}^n$.

In this case $Df(a)$ can be identified with the Jacobian matrix of its partial derivatives and the following significant result holds. The analytic mapping f is one to one near $a \in \Omega$ if and only if f is biholomorphic on a neighborhood $\mathbb{N}(a) \subset \Omega$ if and only if $Df(a)$ is invertible.

This theorem has no analogue if $\mathbb{H}$ is infinite dimensional. In this case we may take $\mathbb{H}$ to be the sequence space l^2. The example below with $x = (x_1, x_2, x_3, \cdots) \in \mathbb{B}$ and

$$f(x) = (x_1^2, x_1^3, x_2^2, x_2^3, x_3^2, x_3^3, \cdots) \tag{1}$$

is analytic and one to one on the ball, has nowhere dense range with $Df(0) = 0$.

In particular f has a functional inverse which is not analytic. In this paper we discuss the strongest result that we know of in the literature to produce analytic invertibility in this setting. This is a result due to Aurich [1] (with an attribution to Abt). His work begins as a study of bifurcation. His tools, which are quite appropriate for his study of the bifurcation theory, are not essential to prove the result of interest to me. The result below appears as a part of a theorem at the end of his paper and I quote only the part of that theorem that pertains to my interest. For each $r \in (0,1]$ denote the open ball with center at the origin and radius r by $\mathbb{B}_r$ with $\mathbb{B}_1 = \mathbb{B}$, and the boundary of $\mathbb{B}_r$ denoted as $\mathbb{S}_r$. The derivatives are bounded operators in $\beta(\mathbb{H})$ and I remind the reader of the class of Fredholm operators.

Definition. An operator $T \in \beta(\mathbb{H})$ is a Fredholm operator if the range of T (written $R(T)$), is closed and the dimension of the null space, $\mathcal{N}(T)$, and the dimension of the co-kernel, $(R(T))^\perp$, are finite.

The notation $(R(T))^\perp$ denotes the orthogonal complement of the closed subspace $R(T)$. The index of T is defined as $i(T) = \dim \mathcal{N}(T) - \dim(R(T))^\perp$.

Date: September 25, 2007.
1991 *Mathematics Subject Classification.* 32K05.

Theorem 1 (Aurich). *Assume f is an analytic and one to one mapping on $\mathbb{B} \to \mathbb{H}$. Assume that for each point $a \in \mathbb{B}$ the operator $Df(a)$ is Fredholm of index zero. Then it follows that f is biholomorphic on $\mathbb{B}$.*

I cast the proof in a Linear Algebra setting and it is of a local nature. The proof is accessible to graduate students that have had a basic course in Functional Analysis.

This type of phenomenon is in some sense typical of the behavior of analytic maps on infinite-dimensional Hilbert spaces. Part of the difficulty in this setting is that the closed unit ball is not compact. In addition to this aspect of topology there is an unsatisfactory aspect to the analyticity of such mappings in the following sense. The analytic function f itself may have interesting properties but we do not know how this affects the behavior of the derivative (e.g., in the example (1) f is one to one yet its derivative at the origin is the zero linear transformation). Conversely, the derivative may possess interesting properties but we have not been able to use these properties to recapture information about the function itself (see Section 3 below). In Theorem 1 strong conditions are imposed on both the function and the derivative.

In the last section I give a short list of problems that I feel are challenging and if solved would flesh out our understanding of the interplay between the local properties of the function and conditions on the derivative. It is difficult to give interesting examples in the infinite-dimensional setting.

I thank Warren Wogen for his advice and interest in this work.

Section 2. Local behavior

The idea is to use a factorization which appears in reference [1].

We begin by stating a special result and then observing its relationship to Theorem 1 above.

Theorem 2. *Suppose that f is an analytic mapping of $\mathbb{B}$ into $\mathbb{H}$ and that $a \in \mathbb{B}$. Let $T = Df(a)$. Set $M = \ker T$ and suppose that $N = (M)^{\perp}$ is the range of T. If M has finite positive dimension, then f is not one to one in any neighborhood of a.*

Proof. We have $\mathbb{H} = M \oplus N$, so relative to this decomposition,

$$T = \begin{pmatrix} O & O \\ O & T_1 \end{pmatrix},$$

where T_1 is invertible from N onto N. We write

$$a = a_1 \oplus a_2 \in M \oplus N,$$

and assume without loss of generality that

$$f(a) = f(a_1 \oplus a_2) = 0 \oplus 0 \in M \oplus N.$$

We can find neighborhoods $U(a_1) \subseteq M$ and $V(a_2) \subseteq N$ so that $U \oplus V$ is a neighborhood of a in B. With P the orthogonal projection of H onto M and $Q = I - P$ the orthogonal projection of H onto N we may write for $x \in B$

$$f(x) = Pf(x) \oplus Qf(x) \equiv u(x) \oplus v(x).$$

Note $u : B \to M$ and $v : B \to N$. We have

$$Du(a) = PDf(a) = PT(a) = 0$$

and

$$Dv(a) = Q(Df(a)) = T_1.$$

By the Implicit Function Theorem for Banach spaces (Ref [2]) applied to $v(a) = (0\oplus 0))$ and $Dv(a) = T_1$, we have neighborhoods $V_1(a_1) \subseteq U \subseteq M$ and $V_2(a_2) \subseteq V \subseteq N$ and an analytic function $h : v_1 \to v_2$, $h(a_1) = a_2$ and $v(m_1 \oplus n_1) = 0$ for $m_1 \in V_1$, $n_1 \in V_2$ if and only if $n_1 = h(m_1)$. This implies that

$$f(m_1 \oplus h(m_1)) = u(m_1 \oplus h(m_1)) \oplus 0,$$

or

$$f(m_1 \oplus h(m_1)) \in M \quad \text{for } m_1 \in V_1.$$

Now for $m_1 \in V_1$, define the mapping $\tilde{u}$ from M into M as

$$\tilde{u}(m) = u(m \oplus h(m)).$$

Since $Du(a)$ is the zero operator, the Chain Rule gives that the operator $D\tilde{u}(a_1)$ is the zero operator on M. Applying the Inverse Mapping Theorem in finite dimensions we have distinct points m_1 and m_2 in V_1 with $\tilde{u}(m_1) = \tilde{u}(m_2)$. But for $m \in V_1$,

$$f(m_1 \oplus h(m_1)) = u(m_1 \oplus h(m_1)) \oplus 0 = \tilde{u}(m_1) \oplus 0$$

and

$$\begin{aligned} f(m_2 \oplus h(m_2)) &= u(m_2 \oplus h(m_2)) \oplus 0 \\ &= \tilde{u}(m_2) \oplus 0 = \tilde{u}(m_2) \oplus 0. \end{aligned}$$

Since the direct sum decomposition is unique, we have $m_1 \oplus h(m_1) \neq m_2 \oplus h(m_2)$ and $f(m_1 \oplus h(m_1)) = f(m_2 \oplus h(m_2))$. Hence, f is not one to one near a_2. □

Theorem 2 yields the proof of Theorem 1. We show that if f is analytic and one to one on $\mathbb{B}$ and if $T = Df(a)$ is Fredholm on index zero at $a \in \mathbb{B}$ then T must be invertible. It then follows by the Inverse Mapping Theorem that f is biholomorphic near a.

Suppose T has index zero and $M = \ker T$ has positive dimension. If N is equal to the range T, then $N^\perp$ has the same dimension as M. So choose a unitary operator U with $U(M) = N^\perp$ and $U(M^\perp) = N$. If $g = Uf$, then g satisfies the hypothesis of Theorem 2, so g is not one to one in any neighborhood of a. Hence, neither if f.

Section 3

There is an interesting example of Heath and Suffridge which does the following. They produce a mapping from the open unit ball in the non-separable Banach space H^∞ onto the space which has an open set in its range; it is one to one and yet the range of f is not an open set. In this case the Banach space is a difficult space to work in and so one might expect more aberrant behavior of analytic mappings because of the difficult structure of such non-separable Banach spaces.

I feel the following problems are worthy of study and positive results or counterexamples would be useful in the study of such mappings. As above $\mathbb{H}$ is a separable Hilbert space.

Q 1. Assume f is analytic and is a one to one mapping from $\mathbb{B} \to \mathbb{H}$ with $f(\mathbb{B})$ an open set. Is f a biholomorphic mapping?

Q 2. Assume f is analytic and is a one to one mapping from $\mathbb{B} \to \mathbb{H}$ with $Df(0) = 0$. Show that range of f can not be locally open at 0.

Q 3. Assume f as above and $Df(x)$ invertible for $0 < \|x\| < r < 1$. Show that f is biholomorphic on $\mathbb{B}$.

References

[1] Aurich, V., Bifurcation of the solutions of Holomorphic Fredholm equations and complex analytic graph theorems, *Nonlinear analysis. Theory, Methods and Applications*, Vol. 6, No. 6, pp 599–613, 1982.

[2] Dieudonne, J., *Foundations of Modern Analysis*, Academic Press, New York and London, 1960.

[3] Heath, L.F. and Suffridge, T.J. Starlike, convex, close to convex, spirallike, and Φ like maps in a commutative Banach algebra with identity, *Transactions of the American Mathematics Society*, Vol. 250, June 1979.

UNC, Chapel Hill, N.C. 27599

Contemporary Mathematics
Volume **454**, 2008

Isometric composition operators on the Bloch space in the polydisk

Joel Cohen and Flavia Colonna

In memory of Alec L. Matheson

Abstract. In this work, we study the holomorphic functions φ mapping the open unit polydisk Δ^n into itself whose induced composition operator $C_\varphi : f \mapsto f \circ \varphi$ is an isometry on the Bloch space in Δ^n. We give some necessary conditions and describe a wide class of such functions by relating their components to symbols of isometric composition operators on the Bloch space of the unit disk Δ, which have a complete characterization.

1. Introduction

Let f be a complex-valued holomorphic function on the unit polydisk

$$\Delta^n = \{z = (z_1, \dots, z_n) \in \mathbb{C}^n : |z_k| < 1, k = 1, \dots, n\}$$

and, for $z \in \Delta^n$ and $u \in \mathbb{C}^n$, let $(\nabla f)(z)u = \sum_{k=1}^n \frac{\partial f}{\partial z_k}(z)u_k$. For $u, v \in \mathbb{C}^n$, denote by

$$H_z(u, \overline{v}) = \sum_{k=1}^n \frac{u_k \overline{v_k}}{(1 - |z_k|^2)^2}$$

the Bergman metric on Δ^n, that is, the positive definite bilinear form which is invariant under biholomorphic transformations of Δ^n. The function f is said to be **Bloch** if $\beta_f = \sup_{z \in \Delta^n} Q_f(z)$ is finite, where

$$Q_f(z) = \sup_{u \in \mathbb{C}^n \setminus \{0\}} \frac{|(\nabla f)(z)u|}{H_z(u, \overline{u})^{1/2}}.$$

Denote by $\mathcal{B}$ the space of all Bloch functions on Δ^n. The map $f \mapsto \beta_f$ is a semi-norm on $\mathcal{B}$, and $\mathcal{B}$ is a Banach space, known as the **Bloch space**, under the norm $\|f\|_{\mathcal{B}} = |f(0)| + \beta_f$. The above definition was given by Timoney in [**20**] on a bounded homogeneous domain (i.e. a bounded domain $D \subset \mathbb{C}^n$ whose group of biholomorphic transformations acts transitively on D) with $H_z(u, \overline{v})$ representing the Bergman metric on the domain, although the notion of Bloch function in higher dimensions was first introduced by K. T. Hahn in [**14**]. For excellent references on

---

1991 *Mathematics Subject Classification.* Primary: 32A18; Secondary: 30D45, 32M15, 47B33.

Key words and phrases. Bloch functions, Bergman metric.

the theory of Bloch functions on a bounded homogeneous domain, see [**20**] and [**21**].

For $n = 1$, the above definition of Bloch semi-norm reduces to the well-known formula $\beta_f = \sup_{z\in\Delta}(1-|z|^2)|f'(z)|$. References on the theory of Bloch functions on the unit disk include [**1**] and [**3**].

A holomorphic function φ mapping Δ^n into itself induces on $\mathcal{B}$ the composition operator $C_\varphi(f) = f \circ \varphi$ which is bounded. Indeed, as a consequence of Theorem 2.12 in [**20**], the operator C_φ is bounded if φ is a holomorphic function mapping a bounded homogeneous domain into itself. The function φ is called the **symbol** of the operator. A useful reference on operator theory in function spaces is [**23**]. As references on composition operators on spaces of analytic functions, see [**19**] and [**10**]. Bounded composition operators on the Bloch space for the polydisk have been studied in [**25**]. In that article, Zhou and Shi obtained the higher-dimensional analogue of the characterization of the compactness of composition operators on the Bloch space proved by Madegan and Matheson for the case of the disk [**16**].

In this paper, we study the problem of characterizing the symbol of the isometric composition operators on the Bloch space of the polydisk. The one-dimensional case, first studied in [**26**], was completely solved in [**7**]. Indeed, the second author showed the following result.

THEOREM 1.1. [**7**] *Given $\varphi : \Delta \to \Delta$ analytic, C_φ is an isometry on $\mathcal{B}$ if and only if $\varphi(0) = 0$ and $\beta_\varphi = 1$.*

In Theorem 2.1 we shall list several statements that are equivalent to the condition $\beta_\varphi = 1$. Using the canonical decomposition of bounded analytic functions on Δ, the symbols φ of the isometric composition operators on $\mathcal{B}$ that are not rotations can be characterized in terms of the zeros and the values at the zeros of the non-vanishing term in their decomposition as a Blaschke product times a nonvanishing analytic function from Δ to $\overline{\Delta}$.

Since for a Blaschke product B with zeros $\{z_k\}_{k\in\mathbb{N}}$, we have

$$(1-|z_k|^2)|B'(z_k)| = \prod_{j\neq k}\left|\frac{z_k - z_j}{1-\overline{z_k}z_j}\right|,$$

non-trivial examples of symbols φ whose corresponding operator C_φ is an isometry on $\mathcal{B}$ include infinite Blaschke products fixing 0 and whose zero set is a sequence $\{z_k\}$ satisfying the condition

$$\lim_{k\to\infty}\prod_{j\neq k}\left|\frac{z_k - z_j}{1-\overline{z_k}z_j}\right| = 1.$$

Sequences satisfying the above condition are called **thin**. In [**11**] it was shown that a sequence $\{z_k\}$ satisfying the condition

$$\lim_{k\to\infty}\frac{1-|z_{k+1}|}{1-|z_k|} = 0$$

is thin. Thus, examples of thin sequences are $\{1 - 1/k!\}$ and $\{1 - k^{-k}\}$.

Thin Blaschke products have the property that the preimage of every $a \in \Delta$ is a thin sequence and are known to be *indestructible*, that is, if B is a Blaschke product whose zeros form a thin sequence, then for all $a \in \Delta$, $L_a \circ B$ is a Blaschke product, where $L_a(z) = \frac{a-z}{1-\overline{a}z}$.

By Theorem 1.1 and Theorem 2.1, the zero sets Z of the symbols of the isometric composition operators that are not rotations form infinite sequences $\{z_k\}_{k\in\mathbb{N}}$ satisfying the condition

$$\limsup_{k\to\infty} \prod_{\zeta\in Z, \zeta\neq z_k} \left|\frac{z_k-\zeta}{1-\overline{z_k}\zeta}\right| = 1.$$

We call these **almost-thin** sequences. Almost-thin sequences form a much wider class than the thin sequences. Indeed, in [**6**] we proved that if C_φ is an isometry on $\mathcal{B}$, where φ is not a rotation, then the preimage of every $a \in \Delta$ under φ is an almost-thin sequence. Let B be a thin Blaschke product whose zeros z_k ($k \in \mathbb{N}$) cluster at a point $\zeta \in \partial\Delta$, and let g be a singular inner function such that $\lim_{z\to\zeta} |g(z)| = 1$. Setting $\varphi = gB$, we see that $\beta_\varphi = 1$ because

$$\lim_{k\to\infty}(1-|z_k|^2)|\varphi'(z_k)| = \lim_{k\to\infty}(1-|z_k|^2)|g(z_k)||B'(z_k)| = 1.$$

Then from Frostman's Theorem (see [**13**], Theorem 6.4) it follows that $\varphi^{-1}(a)$ is an almost thin sequence for all $a \in \Delta$, and is thin only for a set of logarithmic capacity 0.

For domains $D \subset \mathbb{C}^n, D' \subset \mathbb{C}^m$, denote by $H(D, D')$ the class consisting of the holomorphic functions from D to D'.

For a domain D in $\mathbb{C}^n$, let us denote by $\mathrm{Aut}(D)$ the group of biholomorphic transformations of D onto itself, which we call *automorphisms* of D. The automorphisms of Δ^n are the transformations of the form

$$T(z) = (S_1(z_{\tau(1)}), \ldots, S_n(z_{\tau(n)})),\ z = (z_1, \ldots, z_n) \in \Delta^n,$$

where $S_j \in \mathrm{Aut}(\Delta)$, $j = 1, \ldots, n$, and τ is a permutation of $\{1, \ldots, n\}$ (e.g. see [**18**], p.167).

For $\varphi \in H(\Delta^n, \Delta^n)$, let $J\varphi(z) = \left(\frac{\partial\varphi_j}{\partial z_k}(z)\right)_{1\leq j,k\leq n}$ denote the Jacobian matrix of φ at z so that $J\varphi(z)u$ is the usual matrix product where u is viewed as a column vector. Then

$$B_\varphi = \sup_{z\in\Delta^n} \sup_{u\in\mathbb{C}^n\setminus\{0\}} \frac{H_{\varphi(z)}(J\varphi(z)u, \overline{J\varphi(z)u})^{1/2}}{H_z(u,\overline{u})^{1/2}}$$

is finite. Indeed, in [**20**] (proof of Theorem 2.12) it was shown that if $D_1 \subset \mathbb{C}^{n_1}$ and $D_2 \subset \mathbb{C}^{n_2}$ are bounded homogeneous domains, then there is a positive constant c such that for all $\varphi \in H(D_1, D_2)$, $z \in D_1$, and $u \in \mathbb{C}^n$

$$H^{D_2}_{\varphi(z)}(J\varphi(z)u, \overline{J\varphi(z)u}) \leq cH^{D_1}_z(u,\overline{u}),$$

where $H^{D_1}_z$ and $H^{D_2}_{\varphi(z)}$ are the Bergman metrics on D_1 and on D_2 at z and $\varphi(z)$, respectively.

If $\varphi \in \mathrm{Aut}(\Delta^n)$, then for any choice of $z \in \Delta^n$ and $u \in \mathbb{C}^n$

$$H_{\varphi(z)}(J\varphi(z)u, \overline{J\varphi(z)u}) = H_z(u,\overline{u}),$$

which implies $B_\varphi = 1$. Moreover, for any $f \in \mathcal{B}$, $z \in \Delta^n$, we have

$$Q_{f\circ\varphi}(z) \leq \sup_{u\in\mathbb{C}^n\setminus\{0\}} \left(\frac{H_{\varphi(z)}(J\varphi(z)u, \overline{J\varphi(z)u})}{H_z(u,\overline{u})}\right)^{1/2} Q_f(\varphi(z)).$$

Thus

$$Q_{f\circ\varphi}(z) \leq B_\varphi Q_f(\varphi(z)) \tag{1.1}$$

and $Q_{f\circ\varphi}(z) = Q_f(\varphi(z))$ if $\varphi \in \mathrm{Aut}(\Delta^n)$. Consequently, composition operators induced by the automorphisms of Δ^n preserve the Bloch semi-norm. Furthermore, if C_φ preserves the Bloch semi-norm, then $B_\varphi \geq 1$.

When $n = 1$, $H_z(u, \overline{u}) = \frac{|u|^2}{(1-|z|^2)^2}$, and thus

$$B_\varphi = \sup_{z\in\Delta} \frac{(1-|z|^2)|\varphi'(z)|}{1-|\varphi(z)|^2},$$

which is no greater than 1 by the Schwarz-Pick lemma.

REMARK 1.2. In higher dimensions, B_φ may be larger than 1. For example, if $\varphi(z_1, z_2) = (z_1, z_1)$, then $B_\varphi = \sqrt{2}$. But C_φ is not an isometry because it has a non-trivial kernel. Indeed, the function $f(z_1, z_2) = z_1 - z_2$ is Bloch with positive semi-norm, but $C_\varphi(f)$ is identically 0 on Δ^2.

In general, for a holomorphic self-map φ of Δ^n, the value of B_φ can be as large as $\sqrt{n}$. It is precisely this phenomenon that makes the problem of characterizing the isometries C_φ on $\mathcal{B}$ difficult in higher dimensions.

We now give a simpler formula for calculating B_φ. By definition

$$B_\varphi^2 = \sup_{z\in\Delta^n} \sup_{u\in\mathbb{C}^n\setminus\{0\}} \frac{\sum_{k=1}^n \frac{|\sum_{j=1}^n \frac{\partial\varphi_k}{\partial z_j}(z)u_j|^2}{(1-|\varphi_k(z)|^2)^2}}{\sum_{l=1}^n \frac{|u_l|^2}{(1-|z_l|^2)^2}}.$$

Setting $w_j = u_j/(1-|z_j|^2)$, we obtain

$$B_\varphi = \sup_{z\in\Delta^n} \max_{\|w\|=1} \left(\sum_{k=1}^n \frac{\left|\sum_{j=1}^n \frac{\partial\varphi_k}{\partial z_j}(z)(1-|z_j|^2)w_j\right|^2}{(1-|\varphi_k(z)|^2)^2} \right)^{1/2}, \tag{1.2}$$

where $\|w\| = \sqrt{\sum_{j=1}^n |w_j|^2}$.

A bounded domain D in $\mathbb{C}^n$ is said to be **symmetric** if for each $a \in D$ there exists an automorphism S of D for which a is an isolated fixed point and such that $S^{-1} = S$. The polydisk is a bounded symmetric domain, since it is homogeneous and the mapping $S(z) = -z$ is its own inverse and has 0 as an isolated fixed point. Bounded symmetric domains are homogeneous (see [**15**], pp. 170, 301).

E. Cartan [**2**] classified the bounded symmetric domains (up to biholomorphic transformations) into six irreducible classes (all containing 0), four of which consist of large families of domains and two others consisting of a single domain each, known as *exceptional domains.* A finite product of domains of the first four types is known as a *Cartan classical domain.* The polydisk and the unit ball are Cartan classical domains.

In Theorems 2 and 3 of [**5**] we expressed the value c_D of the supremum of the Bloch semi-norms of the bounded holomorphic functions mapping a Cartan classical domain D into Δ in terms of the Bergman metric of the domain at the origin and the corresponding values of the irreducible factors of D. Specifically, we showed that for each irreducible Cartan domain D,

$$c_D = \frac{1}{\inf_{u\in\partial D} H_0^D(u, \overline{u})^{1/2}}.$$

Furthermore, we proved that c_D is the maximum over all $j = 1, \ldots, N$ of c_{D_j}, where $D_1, \ldots, D_N$ are the irreducible factors of D. We also gave in Corollary 1 the specific value of the constant c_D in terms of the class and the dimension of the domain D. This result was later extended to the exceptional domains in [**22**]. In particular, for $D = \Delta^n$, we obtain

$$c_D = \sup\{\beta_f : f \in H(\Delta, \Delta)\} = 1. \tag{1.3}$$

In this article, we present an overview of the many characterizations of the isometric composition operators on the Bloch space in the one-dimensional case (Corollary 2.2). After giving some preliminary results on Bloch functions on the polydisk (Theorem 3.1, Theorem 3.2, and Theorem 3.3), we obtain some necessary conditions on $\varphi = (\varphi_1, \dots, \varphi_n)$ for C_φ to be an isometry on $\mathcal{B}$. Specifically, in Theorem 4.4 we prove that if C_φ is an isometry on $\mathcal{B}$, then $\varphi(0) = 0$ and the components of φ must be linearly independent and have maximal Bloch semi-norm. Moreover, in Theorem 4.5 we show that if C_φ is an isometry on $\mathcal{B}$, then there exist sequences $\{T^{j,(k)}\}_{k\in\mathbb{N}}$ (with $j = 1, \dots, n$) of automorphisms of Δ^n such that $z \mapsto (\varphi_1(T^{1,(k)}(z)), \dots, \varphi_n(T^{n,(k)}(z)))$ converges to the identity of Δ^n.

We use the results presented in section 2 to provide in Theorem 4.10 a wide class of functions φ on Δ^n whose induced composition operator is an isometry on $\mathcal{B}$.

We conclude the paper with a conjecture and some open questions.

2. The one-dimensional case

In this section, we present several characterizations of the isometric composition operators on the Bloch space of the unit disk.

THEOREM 2.1. *For $\varphi \in H(\Delta, \Delta)$, the following statements are equivalent:*

(a) $\beta_\varphi = 1$.

(b) Either $\varphi \in Aut(\Delta)$ or for every $a \in \Delta$ there exists a sequence $\{z_k\}$ in Δ such that $|z_k| \to 1$, $\varphi(z_k) = a$, and

$$\lim_{k\to\infty} \frac{(1 - |z_k|^2)|\varphi'(z_k)|}{1 - |\varphi(z_k)|^2} = 1. \tag{2.1}$$

(c) Either $\varphi \in Aut(\Delta)$ or the zeros of φ form an infinite sequence $\{z_k\}_{k\in\mathbb{N}}$ such that $\limsup_{k\to\infty}(1 - |z_k|^2)|\varphi'(z_k)| = 1$.

(d) Either $\varphi \in Aut(\Delta)$ or $\varphi = gB$ where g is a nonvanishing analytic function mapping Δ into itself or a constant of modulus 1, and B is an infinite Blaschke product whose zero set Z contains a sequence $\{z_k\}$ such that $|g(z_k)| \to 1$ and

$$\limsup_{k\to\infty} \prod_{\zeta\in Z, \zeta\neq z_k} \left|\frac{z_k - \zeta}{1 - \overline{z_k}\zeta}\right| = 1.$$

In particular, the zeros of φ form an almost-thin sequence.

(e) Either $\varphi \in Aut(\Delta)$ or there exists $\{S_k\}_{k\in\mathbb{N}}$ in $Aut(\Delta)$ such that $|S_k(0)| \to 1$ and $\{\varphi \circ S_k\}$ approaches the identity locally uniformly in Δ.

(f) C_φ preserves the Bloch semi-norm on $\mathcal{B}$.

(g) $B_\varphi = 1$.

The fraction on the left-hand side of (2.1) is also known as the *hyperbolic derivative* of φ at z_k. The equivalence of (a) and (b) with the condition $\varphi(z_k) = a$ replaced by the apparently weaker condition $\varphi(z_k) \to a$ was proved in [**17**].

PROOF. The equivalence of (a) and (b) follows from Theorem 2.7 of [**6**]. The equivalence of (a), (c) and (d) was shown in Theorem 3 and Corollary 1 of [**7**]. The equivalence of (a) and (e) was proved in Theorem 2 of [**9**].

Assume (e) holds and show that (f) holds. If $\varphi \in \mathrm{Aut}(\Delta)$, then by the invariance of the Bloch semi-norm under right composition of automorphisms, it follows immediately that $\beta_{f\circ\varphi} = \beta_f$ for each $f \in \mathcal{B}$. So assume there exists $\{S_k\}_{k\in\mathbb{N}}$ in $\mathrm{Aut}(\Delta)$ such that $|S_k(0)| \to 1$ and $\{\varphi \circ S_k\}$ approaches the identity locally uniformly in Δ. Then for each $f \in \mathcal{B}$, $f \circ \varphi \circ S_k$ converges to f locally uniformly in Δ. Theorem 4 in [**7**] says that the correspondence $f \mapsto \beta_f$ is lower-semicontinuous on the Bloch space under the topology of uniform convergence on compact subsets of Δ. Thus, using the invariance of the Bloch semi-norm under right composition by an automorphism and the Schwarz-Pick lemma, we obtain $\beta_f \leq \beta_{f\circ\varphi\circ S_k} = \beta_{f\circ\varphi} \leq \beta_f$, proving (f).

Next, suppose (f) holds. Since the identity has Bloch semi-norm 1, it follows that $\beta_\varphi = 1$, so that (a), (e), and (f) are equivalent statements. Furthermore $\beta_\varphi \leq B_\varphi \leq 1$, proving (g).

Now, assume (g) holds. To complete the proof, it suffices to show that (e) holds. By definition of B_φ, either there exists $w \in \Delta$ such that $(1 - |w|^2)|\varphi'(w)| = 1 - |\varphi(w)|^2$, or there exists a sequence $\{w_k\}$ in Δ such that $|w_k| \to 1$ and

$$\lim_{k\to\infty} \frac{(1 - |w_k|^2)|\varphi'(w_k)|}{1 - |\varphi(w_k)|^2} = 1.$$

In the first case, $\varphi \in \mathrm{Aut}(\Delta)$. In the second case, for each $k \in \mathbb{N}$, let $L_{w_k}(z) = \frac{w_k - z}{1 - \overline{w_k} z}$, for $z \in \Delta$. By a normality argument, we may assume (passing to a subsequence if necessary) that $\{\varphi \circ L_{w_k}\}$ converges to some analytic function $F : \Delta \to \Delta$ such that $|F'(0)| = 1 - |F(0)|^2$. By the Schwarz-Pick lemma, we deduce that $F \in \mathrm{Aut}(\Delta)$. Letting $S_k = L_{w_k} \circ F^{-1}$, it follows that $S_k \in \mathrm{Aut}(\Delta)$, $|S_k(0)| \to 1$, and $\{\varphi \circ S_k\}$ converges locally uniformly to the identity in Δ. □

From Theorem 1.1 and Theorem 2.1 we deduce the following result.

COROLLARY 2.2. *For $\varphi \in H(\Delta, \Delta)$, C_φ is an isometry on $\mathcal{B}$ if and only if $\varphi(0) = 0$ and any of the equivalent conditions (b)-(g) holds. In particular,*

(i) C_φ is an isometry on $\mathcal{B}$ if and only if $\varphi(0) = 0$ and either φ is a rotation or there exists $\{S_k\}_{k\in\mathbb{N}}$ in $\mathrm{Aut}(\Delta)$ such that $|S_k(0)| \to 1$ and $\{\varphi \circ S_k\}$ approaches the identity locally uniformly in Δ.

(ii) C_φ is an isometry on $\mathcal{B}$ if and only if $\varphi(0) = 0$ and $B_\varphi = 1$.

We shall discuss in sections 4 and 5 higher-dimensional versions of (i) and (ii). In [**4**], Cima and Wogen showed that the surjective isometries on the subspace $\mathcal{B}_*$ of the Bloch space of the unit disk consisting of the functions fixing 0 are of the form $T(f) = \lambda(C_\varphi(f) - f(\varphi(0)))$, where λ is a unimodular constant and φ is a conformal automorphism of Δ. Observe that if $\varphi \notin \mathrm{Aut}(\Delta)$ is a function of semi-norm 1, then by the equivalence of statements (a) and (f) in Theorem 2.1, for each unimodular constant λ, the operator $T(f) = \lambda(C_\varphi(f) - f(\varphi(0)))$ preserves the Bloch semi-norm on $\mathcal{B}_*$ because $\beta_{T(f)} = \beta_{f\circ\varphi} = \beta_f$. Since the isometries on $\mathcal{B}_*$ are precisely the functions that preserve the Bloch semi-norm, we deduce that T is a non-surjective isometry on $\mathcal{B}_*$. Part (d) of Theorem 2.1 shows that there is a wide class of non-surjective isometries of this form. We do not know whether there exist non-surjective isometries on $\mathcal{B}_*$ that are not of the above form.

3. Bloch functions on the polydisk

In this section, we give alternate descriptions of the Bloch semi-norm and analyze some properties of Bloch functions. We begin by observing that the Bloch functions are precisely the Lipschitz maps between the polydisk under the distance ρ induced by the Bergman metric and the complex plane under the Euclidean distance. To see this, we recall a very useful result of [**13**] connecting local derivatives to Lipschitz mappings. To put this in context, we look again at the definition of $Q_f(z)$ for $z \in \Delta^n$. If we consider $f : \Delta^n \to \mathbb{C}$ as a map between Riemannian manifolds, then $Q_f(z)$ is exactly the operator norm of $df(z)$, the induced linear transformation on the tangent space at z, with respect to the Bergman metric H_z on Δ^n and the Euclidean metric on $\mathbb{C}$. We recall that β_f is the supremum of $\|df(z)\|$ over all $z \in \Delta^n$. The *dilation* of f, $\mathrm{dil}(f)$, is what might be thought of as the global Lipschitz number of f. In our context,

$$\mathrm{dil}(f) = \sup_{z,w\in\Delta^n, z\neq w} \frac{|f(z) - f(w)|}{\rho(z,w)}.$$

It is the relation between β_f and $\mathrm{dil}(f)$ that we are looking for.

A **length space** is a Riemannian manifold in which the distance between two points is the infimum of the lengths of geodesics connecting the points. Examples include Δ^n as well as $\mathbb{C}$. An example of a space which is not a length space is $\mathbb{C}$ with a closed line segment removed. In Property 1.8 bis of [**13**], Gromov shows that if $f : M \to N$ is a smooth map between Riemannian manifolds and M is a length space, then

$$\mathrm{dil}(f) = \sup_{m\in M} \|df(m)\|.$$

In the present context, we deduce

THEOREM 3.1. *Let $f : \Delta^n \to \mathbb{C}$ be a holomorphic function. Then f is Bloch if and only if f is a Lipschitz map as a function from the polydisk under the Bergman metric and the complex plane under the Euclidean metric. Furthermore*

$$\beta_f = \sup_{z\neq w} \frac{|f(z) - f(w)|}{\rho(z,w)}.$$

A proof that does not use differential geometry for the special case $n = 1$ can be found in [**23**]. For a proof for Bloch functions on the unit ball, see [**24**].

We now prove a convergence theorem for Bloch functions on the polydisk which was proved in [**8**] in the one-dimensional case.

THEOREM 3.2. *Let $\{f_n\}$ be a sequence of Bloch functions converging locally uniformly to some holomorphic function f. If the sequence $\{\beta_{f_n}\}$ is bounded, then f is a Bloch function and*

$$\beta_f \leq \liminf_{n\to\infty} \beta_{f_n}.$$

That is, the function $f \mapsto \beta_f$ is lower semi-continuous on $\mathcal{B}$ with respect to the topology of uniform convergence on compact subsets of Δ^n.

PROOF. Since $\{\beta_{f_n}\}$ is bounded, $B = \liminf_{n\to\infty} \beta_{f_n}$ exists and is a non-negative number. Let $\{n_k\}$ be a sequence of positive integers such that $B = \lim_{k\to\infty} \beta_{f_{n_k}}$. Let $z, w \in \Delta^n$ and fix $\epsilon > 0$. Choose $\nu \in \mathbb{N}$ such that $|f_{n_k}(z) - f(z)| < \epsilon/2$, $|f_{n_k}(w) - f(w)| < \epsilon/2$, and $\beta_{f_{n_k}} < B + \epsilon$ for all $k \geq \nu$. Then

$$|f(z) - f(w)| < \epsilon + \beta_{f_{n_k}} \rho(z,w) < \epsilon(1 + \rho(z,w)) + B\rho(z,w).$$

Letting $\epsilon \to 0$, we obtain $|f(z) - f(w)| \leq B\rho(z, w)$. By Theorem 3.1, f is a Bloch function, and $\beta_f \leq B$. □

We now give another description of the Bloch semi-norm.

THEOREM 3.3. *Let $f \in \mathcal{B}$. Then*

$$\beta_f = \sup_{z \in \Delta^n} \left\| \left(\frac{\partial f}{\partial z_1}(z)(1 - |z_1|^2), \ldots, \frac{\partial f}{\partial z_n}(z)(1 - |z_n|^2) \right) \right\|.$$

For the proof we need the following lemma.

LEMMA 3.4. *Let $w_k \in \mathbb{C}$ and let $a_k > 0$ for $k = 1, \ldots, n$. Then*

$$\max_{\|u\|=1} \frac{\left|\sum_{k=1}^n w_k u_k\right|^2}{\sum_{k=1}^n a_k |u_k|^2} = \sum_{k=1}^n \frac{|w_k|^2}{a_k}.$$

PROOF. The above equality is obvious if each $w_k = 0$. So assume that at least one of the w_k is nonzero. Choose the arguments of $u_1, \ldots, u_n$ so that $|\sum_{k=1}^n w_k u_k| = \sum_{k=1}^n |w_k||u_k|$. Next observe that

$$\frac{\left(\sum_{k=1}^n |w_k||u_k|\right)^2}{\sum_{k=1}^n a_k |u_k|^2} \leq \sum_{k=1}^n \frac{|w_k|^2}{a_k},$$

and equality is attained for $|u_j| = \frac{|w_j|/a_j}{\left(\sum_{k=1}^n (|w_k|/a_k)^2\right)^{1/2}}$, for each $j = 1, \ldots, n$. □

Proof of Theorem 3.3. Fixing $z \in \Delta^n$ and applying Lemma 3.4, we obtain

$$\begin{aligned} Q_f(z)^2 &= \sup_{u \in \mathbb{C}^n \setminus \{0\}} \frac{\left|\sum_{k=1}^n \frac{\partial f}{\partial z_k}(z) u_k\right|^2}{H_z(u, \overline{u})} = \max_{\|u\|=1} \frac{\left|\sum_{k=1}^n \frac{\partial f}{\partial z_k}(z) u_k\right|^2}{\sum_{k=1}^n \frac{|u_k|^2}{(1-|z_k|^2)^2}} \\ &= \sum_{k=1}^n \left|\frac{\partial f}{\partial z_k}(z)\right|^2 (1 - |z_k|^2)^2. \end{aligned}$$

□

NOTATION 3.5. *In the remainder of the paper, to avoid confusion, sequences in higher dimensions will contain a superscript rather than the conventional subscript, which will be reserved for the components.*

We recall a result that will be used in the next section to prove a convergence theorem for holomorphic functions from the polydisk into itself.

THEOREM 3.6. (Corollary 4 of [**5**]) *Let $D = D_1 \times \cdots \times D_N$ be a Cartan classical domain with $D_1, \ldots, D_N$ irreducible, where least one of the factors is Δ. Let $f \in H(D, \Delta)$ with $\beta_f = c_D$. Then there exists a sequence $\{S^k\}_{k \in \mathbb{N}}$ of automorphisms of D and an integer $m = 1, \ldots, N$ such that $D_m = \Delta$ and $\{f \circ S^k\}$ converges locally uniformly in D to the projection map $z \mapsto z_m$.*

4. Holomorphic self maps of Δ^n inducing an isometry on $\mathcal{B}$

The polydisk, viewed as a Riemannian manifold under the structure induced by the Bergman metric, is a length space. Thus, applying [**13**], Prop. 1.8 bis as in Theorem 3.1, we obtain the following result.

PROPOSITION 4.1. *Let $\varphi \in H(\Delta^n, \Delta^n)$. Then φ is Lipschitz as a function from (Δ^n, ρ) into itself and B_φ is the Lipschitz number of φ. That is*

$$B_\varphi = \sup_{z \neq w} \frac{\rho(\varphi(z), \varphi(w))}{\rho(z, w)}.$$

Thus, B_φ can be interpreted as the *Bergman dilation* of φ.

From Proposition 4.1 and the invariance of the Bergman distance under the action of $\mathrm{Aut}(\Delta^n)$, we deduce

COROLLARY 4.2. *The correspondence $\varphi \in H(\Delta^n, \Delta^n) \mapsto B_\varphi$ is invariant under left and right composition of automorphisms of Δ^n.*

From Proposition 4.1, we also derive the lower semi-continuity of the map $\varphi \mapsto B_\varphi$ on the space $H(\Delta^n, \Delta^n)$.

COROLLARY 4.3. *Let $\{\varphi^k\}_{k\in\mathbb{N}}$ be a sequence in $H(\Delta^n, \Delta^n)$ converging locally uniformly to a nonconstant holomorphic function φ. Then $B_\varphi \leq \liminf_{k\to\infty} B_{\varphi^k}$.*

The proof is analogous to that of Theorem 3.2, where we replace the Euclidean metric with the Bergman metric.

We now give some necessary conditions for φ to be the symbol of an isometric composition operator on $\mathcal{B}$.

THEOREM 4.4. *Let $\varphi \in H(\Delta^n, \Delta^n)$ such that C_φ is an isometry on $\mathcal{B}$. Then $\varphi(0) = 0$ and for each $j = 1, \ldots, n$, $\beta_{\varphi_j} = 1$. Furthermore, the components of φ are linearly independent.*

PROOF. Given φ as in the hypothesis, let $\varphi(0) = a$ and write $a = (a_1, \ldots, a_n)$. Fix $j = 1, \ldots, n$ and set $\psi_j(z) = \frac{a_j - z_j}{1 - \overline{a_j} z_j}$. Then for each $z \in \Delta^n$, and each $u \in \mathbb{C}^n$ with $\|u\| = 1$, we have $\sum_{k=1}^n \frac{|u_k|^2}{(1-|z_k|^2)^2} \geq \frac{|u_j|^2}{(1-|z_j|^2)^2}$ with equality holding for $|u_j| = 1$ and $u_k = 0$ for all $k \neq j$. Thus

$$\max_{\|u\|=1} \frac{|u_j|}{H_z(u, \overline{u})^{1/2}} = \max_{\|u\|=1} \frac{|u_j|}{\left(\sum_{k=1}^n \frac{|u_k|^2}{(1-|z_k|^2)^2}\right)^{1/2}} = 1 - |z_j|^2.$$

Consequently

$$\begin{aligned}
\|\psi_j\|_\mathcal{B} &= |a_j| + \sup_{z\in\Delta^n} \max_{\|u\|=1} \frac{|(\nabla\psi_j)(z)u|}{H_z(u,\overline{u})^{1/2}} \\
&= |a_j| + \sup_{z\in\Delta^n} \max_{\|u\|=1} \frac{(1-|a_j|^2)|u_j|}{|1-\overline{a_j}z_j|^2 H_z(u,\overline{u})^{1/2}} \\
&= |a_j| + \sup_{z\in\Delta^n} \frac{(1-|a_j|^2)(1-|z_j|^2)}{|1-\overline{a_j}z_j|^2} \\
&= |a_j| + \sup_{z\in\Delta^n} \left(1 - \left|\frac{a_j - z_j}{1-\overline{a_j}z_j}\right|^2\right) \\
&= |a_j| + 1.
\end{aligned}$$

Since C_φ is an isometry, $\|\psi_j \circ \varphi\|_\mathcal{B} = \|\psi_j\|_\mathcal{B}$. Moreover, since $\psi_j(a) = 0$, by (1.3) we obtain $\|\psi_j \circ \varphi\|_\mathcal{B} = |\psi_j(a)| + \beta_{\psi_j\circ\varphi} = \beta_{\psi_j\circ\varphi} \leq 1$. Hence $a_j = 0$ and $\beta_{\psi_j\circ\varphi} = \beta_{\psi_j} = 1$. Consequently, $\psi_j(z) = -z_j$ and since j was arbitrary, $\varphi(0) = 0$.

In addition, $\nabla(\psi_j \circ \varphi) = -\nabla\varphi_j$, so

$$1 = \beta_{\psi_j \circ \varphi} = \sup_{z\in\Delta^n}\sup_{u\neq 0} \frac{|\nabla(\psi_j \circ \varphi)(z)u|}{H_z(u,\overline{u})^{1/2}} = \sup_{z\in\Delta^n}\max_{\|u\|=1} \frac{|\nabla(\varphi_j)(z)u|}{H_z(u,\overline{u})^{1/2}} = \beta_{\varphi_j}.$$

Next, assume there exist $k = 1,\dots,n$ and constants $a_j \in \mathbb{C}$ for $j \neq k$ such that $\varphi_k = \sum_{j\neq k} a_j\varphi_j$. Then the function f defined by $f(z) = z_k - \sum_{j\neq k} a_j z_j$ is bounded and hence Bloch and has nonzero norm, but $f \circ \varphi$ is identically zero. Thus C_φ is not an isometry on $\mathcal{B}$. □

Theorem 4.5 below is an attempt at getting a higher-dimensional analogue of (i) in Corollary 2.2.

THEOREM 4.5. *Let φ be a holomorphic function from Δ^n into itself such that C_φ is an isometry on $\mathcal{B}$. Then there exist n sequences $\{T^{j,(k)}\}_{k\in\mathbb{N}}$ $(j = 1,\dots,n)$ in $\mathrm{Aut}(\Delta^n)$ such that $(\varphi_1 \circ T^{1,(k)},\dots,\varphi_n \circ T^{n,(k)})$ converges uniformly on compact subsets to the identity of Δ^n.*

PROOF. By Theorem 4.4, $\varphi_j(0) = 0$ and $\beta_{\varphi_j} = 1$ for each $j = 1,\dots,n$. By Theorem 3.6, for each $j = 1,\dots,n$ there exists a sequence $\{S^{j,(k)}\}_{k\in\mathbb{N}}$ of automorphisms of Δ^n and $m_j \in \{1,\dots,n\}$ such that $\varphi_j \circ S^{j,(k)}$ converges locally uniformly to the projection map $z \mapsto z_{m_j}$. Let $\Psi^j : \Delta^n \to \Delta^n$ be defined by $\Psi^j = (\Psi^j_1,\dots,\Psi^j_n)$, where for each $h = 1,\dots,n$, and each $z \in \Delta^n$

$$\Psi^j_h(z) = \begin{cases} z_h & \text{if } h \neq j, m_j, \\ z_{m_j} & \text{if } h = j, \\ z_j & \text{if } h = m_j, \end{cases}$$

and set $T^{j,(k)} = S^{j,(k)} \circ \Psi^j \in \mathrm{Aut}(\Delta^n)$. Then the sequence $\{\varphi_j \circ T^{j,(k)}\}$ converges locally uniformly to the projection function $z \mapsto z_j$. Consequently, the sequence $\{(\varphi_1 \circ T^{1,(k)},\dots,\varphi_n \circ T^{n,(k)})\}$ converges locally uniformly to the identity in Δ^n. □

The converse of Theorem 4.5 is false, as the following simple example shows.

EXAMPLE 4.6. For $z_1, z_2 \in \Delta$, $\varphi(z_1,z_2) = (z_1,z_1)$ and let T^1 and T^2 be the automorphisms of Δ^2 defined by $T^1(z_1,z_2) = (z_1,z_2)$ and $T^2(z_1,z_2) = (z_2,z_1)$. Then $(\varphi_1 \circ T^1(z_1,z_2), \varphi_2 \circ T^2(z_1,z_2)) = (z_1,z_2)$, yet, as observed in Remark 1.2, C_φ is not an isometry on $\mathcal{B}$.

REMARK 4.7. Assume $\varphi \in H(\Delta^n,\Delta^n)$ such that $\varphi(0) = 0$. To verify whether φ is an isometry on $\mathcal{B}$ it suffices to check whether C_φ preserves the Bloch semi-norm. Indeed, if for all $f \in \mathcal{B}$, $\beta_{f\circ\varphi} = \beta_f$, then $\|f \circ \varphi\|_{\mathcal{B}} = |f(0)| + \beta_{f\circ\varphi} = \|f\|_{\mathcal{B}}$.

REMARK 4.8. Assume $\varphi \in H(\Delta^n,\Delta^n)$ such that $\varphi(0) = 0$. If there exists a sequence $\{T^k\}_{k\in\mathbb{N}}$ in $\mathrm{Aut}(\Delta^n)$ such that $\{\varphi \circ T^k\}$ converges to the identity in Δ^n, then by Corollary 4.3 and Corollary 4.2, $1 \leq B_{\varphi\circ T^k} = B_\varphi$. Moreover, the components of φ are linearly independent and, by the lower semicontinuity of $f \mapsto \beta_f$, they have semi-norm 1 since the projection maps have this property. Thus all the necessary conditions in Theorem 4.4 are satisfied. In particular, for each $f \in \mathcal{B}$, the sequence $\{f\circ\varphi\circ T^k\}$ converges to f, so by the lower semicontinuity and the invariance of the Bloch semi-norm under right composition of automorphisms $\beta_f \leq \beta_{f\circ\varphi}$. On the other hand, $\beta_{f\circ\varphi} \leq B_\varphi\beta_f$, and so to obtain equality of the semi-norms β_f and $\beta_{f\circ\varphi}$, we need the additional hypothesis $B_\varphi = 1$. Consequently, in dimension $n \geq 2$ the above convergence condition alone is not sufficient to guarantee that C_φ is an isometry on $\mathcal{B}$.

The following lemma is straightforward and so we omit the proof.

LEMMA 4.9. *Given nonnegative numbers $r_1, \ldots, r_n$,*

$$\max_{\|w\|=1} \sum_{k=1}^{n} r_k |w_k|^2 = \max\{r_1, \ldots, r_n\}.$$

The following result shows that the class of functions inducing an isometric composition operator on the Bloch space is quite large. In particular, we obtain non-trivial examples of isometric composition operators on $\mathcal{B}$ by considering functions whose components depend on a distinct single variable and constitute almost-thin Blaschke products fixing 0.

THEOREM 4.10. *For $z \in \Delta^n$ and τ permutation of $\{1, \ldots, n\}$, let $\varphi(z) = (h_1(z_{\tau(1)}), \ldots, h_n(z_{\tau(n)}))$, where for each $j = 1, \ldots, n$, $h_j \in H(\Delta, \Delta)$ such that $h_j(0) = 0$, $\beta_{h_j} = 1$. Then C_φ is an isometry on $\mathcal{B}$.*

PROOF. For simplicity of notation, we shall assume that $\tau(j) = j$ for all $j = 1, \ldots, n$. Let us set aside the trivial case when each $h_j \in \mathrm{Aut}(\Delta)$. Observe that $\varphi_j(z) = h_j(z_j)$ has semi-norm 1 since Theorem 3.3 implies that

$$\beta_{\varphi_j} = \sup_{z \in \Delta^n} \left(\sum_{k=1}^{n} \left| \frac{\partial \varphi_j(z)}{\partial z_k} \right|^2 (1 - |z_k|^2)^2 \right)^{1/2} = \sup_{z_j \in \Delta} |h_j'(z_j)|(1 - |z_j|^2) = 1.$$

Next we show that $B_\varphi = 1$. Recalling (1.2) and using Lemma 4.9 we deduce

$$\begin{aligned} B_\varphi &= \sup_{z \in \Delta^n} \max_{\|w\|=1} \left(\sum_{k=1}^{n} \frac{|\sum_{j=1}^{n} \frac{\partial \varphi_k}{\partial z_j}(z)(1 - |z_j|^2) w_j|^2}{(1 - |\varphi_k(z)|^2)^2} \right)^{1/2} \\ &= \sup_{z \in \Delta^n} \max_{\|w\|=1} \left(\sum_{k=1}^{n} \frac{|h_k'(z_k)|^2 (1 - |z_k|^2)^2 |w_k|^2}{(1 - |h_k(z_k)|^2)^2} \right)^{1/2} \\ &= \sup_{z \in \Delta^n} \max_{1 \le k \le n} \frac{|h_k'(z_k)|(1 - |z_k|^2)}{1 - |h_k(z_k)|^2} = 1. \end{aligned}$$

By Remark 4.7, to prove that C_φ is an isometry we need to show that it preserves the Bloch semi-norm, and hence, dividing a (nonconstant) Bloch function by its semi-norm, it suffices to show that $\beta_{f \circ \varphi} = 1$ for each $f \in \mathcal{B}$ having semi-norm 1. From (1.1) we deduce that

$$\beta_{f \circ \varphi} \le \beta_f = 1. \tag{4.1}$$

On the other hand, by Theorem 3.3, there exists a sequence $\{a^m\}$ in Δ^n such that

$$\sum_{k=1}^{n} \left| \frac{\partial f}{\partial z_k}(a^m) \right|^2 (1 - |a_k^m|^2)^2 \to 1$$

as $m \to \infty$. By Theorem 2.1 (b), for each $k = 1, \ldots, n$ corresponding to each a_k^m, there exists a sequence $\{(z_k^m)_\nu\}$ in Δ such that $h_k((z_k^m)_\nu) = a_k^m$ and

$$\lim_{\nu \to \infty} \frac{(1 - |(z_k^m)_\nu|^2)|h_k'((z_k^m)_\nu)|}{1 - |h_k((z_k^m)_\nu)|^2} = 1.$$

Let $(z^m)_\nu = ((z_1^m)_\nu, \ldots, (z_n^m)_\nu)$. Then $\varphi((z^m)_\nu) = (h_1((z_1^m)_\nu), \ldots, h_n((z_n^m)_\nu)) = a^m$ and

$$Q_{f\circ\varphi}((z^m)_\nu) = \left(\sum_{k=1}^{n} \left|\frac{\partial f}{\partial z_k}(a^m)\right|^2 (1-|a_k^m|^2)^2\right)^{1/2} \frac{(1-|(z_k^m)_\nu|^2)|h_k'((z_k^m)_\nu)|}{1-|h_k((z_k^m)_\nu)|^2} \to 1$$

as $m, \nu \to \infty$. Thus $\beta_{f\circ\varphi} = \sup_{z\in\Delta^n} Q_{f\circ\varphi}(z) \geq 1$. From (4.1) it follows that $\beta_{f\circ\varphi} = 1$, proving the result. □

5. Open questions

We propose the following conjecture as a generalization of the one-dimensional case (Theorem 1.1 and Corollary 2.2 (ii)).

CONJECTURE 1. *For $\varphi \in H(\Delta^n, \Delta^n)$, C_φ is an isometry on $\mathcal{B}$ if and only if $\varphi(0) = 0$, $\beta_{\varphi_j} = 1$ for each $j = 1, \ldots, n$, and $B_\varphi = 1$.*

We end the paper with the following open questions.

1. If C_φ is an isometry on $\mathcal{B}$, does there exist a sequence $\{T_k\}$ in $\mathrm{Aut}(\Delta^n)$ such that $\{\varphi \circ T_k\}$ converges to the identity? By Corollary 2.2 (i), this is true for $n = 1$.

2. If C_φ is an isometry on $\mathcal{B}$, must the components φ_j depend only on one single complex variable k_j such that $\{k_1, \ldots, k_n\}$ is a permutation of $\{1, \ldots, n\}$, as in Theorem 4.10? With this choice of function φ, a sequence of automorphisms of Δ^n yielding a positive answer to question 1. exists as a consequence of Corollary 2.2(i). Furthermore, a positive answer to question 2. would imply that the symbols of the isometric composition operators are completely characterized by the functions φ described in Theorem 4.10.

3. In the one-dimensional case, it is well known that

$$\rho(z, w) = \sup\{|f(z) - f(w)| : \beta_f \leq 1\} \tag{5.1}$$

for any $z, w \in \Delta$ (see [**23**], Theorem 5.1.7). From Theorem 3.1 it follows immediately that $\sup\{|f(z) - f(w)| : \beta_f \leq 1\} \leq \rho(z, w)$ for all $z, w \in \Delta^n$. Does the opposite inequality hold for $n \geq 2$? Formula (5.1) also holds if the domain is the unit ball of $\mathbb{C}^n$ (see [**24**], Theorem 3.9).

References

[1] J. M. Anderson, J. Clunie, Ch. Pommerenke, *On Bloch functions and normal functions*, J. Reine Angew. Math. **279**(1974), 12–37.

[2] E. Cartan, *Sur les domains bournés de l'espace de n variable complexes*, Abh. Math. Sem. Univ. Hamburgh **11** (1935), 116–162.

[3] J. Cima, *The basic properties of Bloch functions*, Internat. J. Math. & Math. Sci. (2)**3** (1979), 369–413.

[4] J. Cima, W. Wogen, *On isometries of the Bloch spaces*, Illinois J. Math. **24** (1980), 313–316.

[5] J. M. Cohen, F. Colonna, *Bounded holomorphic functions on bounded symmetric domains*, Trans. Amer. Math. Soc. (1)**343** (1994), pp. 135–156.

[6] J. M. Cohen, F. Colonna, *Preimages of one-point sets of Bloch and normal functions*, Mediterr. J. Math. **3**(2006), 513–532.

[7] F. Colonna, *Characterisation of the isometric composition operators on the Bloch space*, Bull. Austral. Math. Soc., **72** (2005), pp. 283–290.

[8] F. Colonna, *Bloch and normal functions and their relation*, Rend. Circ. Mat. Palermo, Ser. II **XXXVIII** (1989), pp. 161–180.

[9] F. Colonna, *The Bloch constant of bounded analytic functions*, J. London Math. Soc. (2)**36** (1987), pp. 95–101.

[10] C. Cowen, B. MacCluer, *Composition Operators on Spaces of Analytic Functions*, Studies in Advanced Mathematics, CRC Press, Boca Raton (1995).

[11] P. Gorkin, R. Mortini, *Universal Blaschke products*, Math. Proc. Camb. Phil. Soc., **136** (2004), 175–184.

[12] P. Gorkin, R. Mortini, *Value distribution of interpolating Blaschke products*, J. London Math. Soc., (2)**72** (2005), 151–168.

[13] M. Gromov, *Structures Métriques pour les Variétés Riemaniennes*, Cedic/Fernand Nathan, Paris, 1981.

[14] K. T. Hahn, *Holomorphic mappings of the hyperbolic space into the complex Euclidean space and the Bloch theorem*, Canad. J. Math. **27** (1975), 446–458.

[15] S. Helgason, *Differential Geometry and Symmetric Spaces*, Academic Press, New York-London, 1962.

[16] K. Madigan, A. Matheson, *Compact composition operators on the Bloch space*, Trans. Amer. Math. Soc., (7)**347**, 1995.

[17] M. J. Martín, D. Vukotić, *Isometries of the Bloch space among the composition operators*, Bull. London Math. Soc. (to appear).

[18] W. Rudin, *Function Theory in Polydiscs*, W. A. Benjamin, Inc., New York, 1969.

[19] J. H. Shapiro, *Composition operators and classical function theory*, Springer, New York, 1993.

[20] R. M. Timoney, *Bloch functions in several complex variables, I*, Bull. London Math. Soc. **12** (1980), 241–267.

[21] R. M. Timoney, *Bloch functions in several complex variables, II*, J. Reine Angew. Math. **319** (1980), 1–22.

[22] G. Zhang, *Bloch constants of bounded symmetric domains*, Trans. Amer. Math. Soc. **349**(1997), 2941–2949.

[23] K. Zhu, *Operator Theory in Function Spaces*, Marcel Dekker, New York, 1990.

[24] K. Zhu, *Spaces of Holomorphic Functions in the Unit Ball*, Springer, New York, 2005.

[25] Z. Zhou, J. Shi, *Composition operators on the Bloch space in polydisks*, Complex Variables **46** (2001), 73–88.

[26] C. Xiong, *Norm of composition operators on the Bloch space*, Bull. Austral. Math. Soc. **70** (2004), pp. 293–299.

University of Maryland, College Park, Maryland
E-mail address: jmc@math.umd.edu

George Mason University, Fairfax, Virginia
E-mail address: fcolonna@gmu.edu

Contemporary Mathematics
Volume **454**, 2008

Pluripolarity of Manifolds

Oleg Eroshkin

1. Introduction

A set $E \subset \mathbb{C}^n$ is called *pluripolar* if there exists a non-constant plurisubharmonic function ϕ such that $\phi \equiv -\infty$ on E. Pluripolar sets form a natural category of "small" sets in complex analysis. Pluripolar sets are polar, so they have Lebesgue measure zero, but there are no simple criteria to determine pluripolarity. In this paper we discuss the conditions that ensure pluripolarity for smooth manifolds. This problem has a long history. S. Pinchuk [**16**], using Bishop's "gluing disks" method, proved that a generic manifold of class C^3 is non-pluripolar. Recall that the manifold $M \subset \mathbb{C}^n$ is called *generic at a point* $p \in M$, if the tangent space T_pM is not contained in a proper complex subspace of $\mathbb{C}^n$. A. Sadullaev [**17**], using the same method proved that a subset of positive measure of a generic manifold of class C^3 is non-pluripolar.

In the opposite direction E. Bedford [**4**] showed that a real-analytic nowhere generic manifold is pluripolar. For some applications to harmonic analysis the condition of real-analyticity is too restrictive. However, the result does not hold for merely smooth manifolds. K. Diederich and J. E. Fornaess [**7**] found an example of a non-pluripolar smooth curve in $\mathbb{C}^2$. They construct a function $f \in C^\infty[0,1]$ such that the graph of this function is not pluripolar. In this example the derivatives $f^{(k)}$ grow very fast as $k \to \infty$.

Recently, D. Coman, N. Levenberg and E. A. Poletsky [**6**] proved that curves of Gevrey class G^s, $s < n+1$ in $\mathbb{C}^n$ are pluripolar. We generalize this result to higher dimensional manifolds. Recall that the submanifold $M \subset \mathbb{C}^n$ is called *totally real* if for every $p \in M$ the tangent space T_pM contains no complex line.

THEOREM 1.1. *Let $M \subset \mathbb{C}^n$ be a totally real submanifold of Gevrey class G^s. If $\dim M = m$ and $ms < n$, then M is pluripolar.*

In fact we prove a stronger result. It follows from Theorem 2.1 in [**1**] that a compact set $X \subset \mathbb{C}^n$ is pluripolar if and only if for any bounded domain D containing X and $\epsilon > 0$ there exists polynomial P such that

$$\sup\{|P(z)| : z \in D\} \geq 1$$

2000 *Mathematics Subject Classification.* Primary 32U20.

and

$$\sup\{|P(z)| : z \in X\} \le \epsilon^{\deg P} .$$

THEOREM 1.2. *Let $M \subset \mathbb{C}^n$ be a totally real submanifold of Gevrey class G^s. Let X be a compact subset of M. If $\dim M = m$ and $ms < n$, then for every $h < \frac{n}{ms}$ and every $N > N_0 = N_0(h)$ there exists a non-constant polynomial $P \in \mathbb{Z}[z_1, z_2, \ldots, z_n]$, $\deg P \le N$ with coefficients bounded by $exp(N^h)$, such that*

$$\sup\{|P(z)| : z \in X\} < \exp(-N^h) . \tag{1.1}$$

This result is similar to the construction of an auxiliary function in transcendental number theory (cf.[**18**] Proposition 4.10).

The Theorem 1.1 gives some information about polynomially convex hulls of manifolds of Gevrey class. Recall, that the *polynomially convex hull* $\widehat{X}$ of X consists of all $z \in \mathbb{C}^m$ such that

$$|P(z)| \le \sup_{\zeta \in X} |P(\zeta)| ,$$

for all polynomials P. It is well known, that the polynomially convex hull of a pluripolar compact set is pluripolar (this follows immediately from Theorem 4.3.4 in [**11**]).

We also introduce the notion of *Kolmogorov dimension* for a compact subset $X \subset \mathbb{C}^n$ (denoted $\mathcal{K}$-dim X) with the following properties.

(1) $0 \le \mathcal{K}\text{-dim } X \le n$.
(2) $\mathcal{K}\text{-dim } \widehat{X} = \mathcal{K}\text{-dim } X$.
(3)
$$\mathcal{K}\text{-dim } \bigcup_{j=1}^{m} X_j = \max\{\mathcal{K}\text{-dim } X_j : j = 1, \ldots, m\} .$$
(4) If $D \subset \mathbb{C}^n$ is a domain, $X \subset D$ and $\phi : D \to \mathbb{C}^k$ is a holomorphic map, then $\mathcal{K}\text{-dim } \phi(X) \le \mathcal{K}\text{-dim } X$.
(5) If $\mathcal{K}\text{-dim } X < n$, then X is pluripolar.

The main result of this paper is the following estimate of the Kolmogorov dimension of totally real submanifolds of Gevrey class.

THEOREM 1.3. *Let $M \subset \mathbb{C}^n$ be a totally real submanifold of Gevrey class G^s. Let X be a compact subset of M. If $\dim M = m$ then $\mathcal{K}$-dim $X \le ms$.*

REMARK 1.4. This estimate is sharp. The similar estimates hold for more general class of CR-manifolds. These issues will be addressed in the forthcoming paper.

In the next section we recall the definition and basic properties of functions of Gevrey class. The notion of Kolmogorov dimension of X is defined in terms of ε-entropy of traces on X of bounded holomorphic functions. The definition and basic properties of ε-entropy are given in Section 3. In Section 4 we discuss the notion of Kolmogorov dimension. The proof of Theorem 1.3 is sketched in Section 5.

The author wishes to thank the referee for useful comments and numerous suggestions.

2. Gevrey Class

We need to introduce some notation first. For a multi-indices $\alpha = (\alpha_1, \alpha_2, \dots, \alpha_m)$, $\beta = (\beta_1, \beta_2, \dots, \beta_m)$ we define $|\alpha| = \sum_{j=1}^m \alpha_j$, $\alpha! = \prod_{j=1}^m \alpha_j!$, and

$$\binom{\alpha}{\beta} = \frac{\alpha!}{(\alpha-\beta)!\beta!} .$$

For an integer k we define $\alpha + k = (\alpha_1 + k, \alpha_2 + k, \dots, \alpha_m + k)$. For a point $x \in \mathbb{R}^m$ we define $x^\alpha = \prod_{j=1}^m x_j^{\alpha_j}$. If $f \in C^\infty(\mathbb{R}^m)$ we denote

$$\partial^\alpha f = \frac{\partial^{|\alpha|}}{\partial^{\alpha_1} x_1 \dots \partial^{\alpha_m} x_m} f .$$

Let U be an open set in $\mathbb{R}^m$ and $s \geq 1$. A function $f \in C^\infty(U)$ is said to belong to Gevrey class $G^s(U)$ if for every compact $K \subset U$ there exists a constant $C_K > 0$ such that

$$\sup_{x \in K} |\partial^\alpha f(x)| \leq C_K^{|\alpha|+1} (\alpha!)^s , \tag{2.1}$$

for every multi-index α. The class G^s forms an algebra. The Gevrey class G^s is closed with respect to composition and the Implicit Function Theorem holds for G^s [**13**], thus one may define manifolds (and submanifolds) of Gevrey class G^s in the usual way.

3. The notion of ε-entropy

Let (E, ρ) be a totally bounded metric space. A family of sets $\{C_j\}$ of diameter not greater than 2ε is called an *ε-covering* of E if $E \subseteq \bigcup C_j$. Let $N_\varepsilon(E)$ be the smallest cardinality of the ε-covering.

A set $Y \subseteq E$ is called *ε-distinguishable* if the distance between any two points in Y is greater than ε: $\rho(x, y) > \varepsilon$ for all $x, y \in Y$, $x \neq y$. Let $M_\varepsilon(E)$ be the largest cardinality of an ε-distinguishable set.

For a nonempty totally bounded set E the natural logarithm

$$\mathcal{H}_\varepsilon(E) = \log N_\varepsilon(E)$$

is called the *ε-entropy.*

The notion of ε-entropy was introduced by A. N. Kolmogorov in the 1950's. Kolmogorov was motivated by Vitushkin's work on Hilbert's 13th problem and Shannon's information theory. Note that Kolmogorov's original definition (see [**12**]) is slightly different from ours (he used the logarithm to base 2). Here we follow [**14**].

We will need some basic properties of the ε-entropy.

LEMMA 3.1. *(see [**12**], Theorem IV) For each totally bounded space E and each $\varepsilon > 0$*

$$M_{2\varepsilon}(E) \leq N_\varepsilon(E) \leq M_\varepsilon(E) \tag{3.1}$$

LEMMA 3.2. *Let $\{(E_j, \rho_j) : j = 1, 2, \dots, k\}$ be a family of totally bounded metric spaces. Let (E, ρ) be a Cartesian product with a sup-metric, i.e.*

$$E = E_1 \times E_2 \times \dots \times E_k ,$$

$$\rho((x_1, x_2, \dots, x_k), (y_1, y_2, \dots, y_k)) = max_j \rho_j(x_j, y_j) .$$

Then

$$\mathcal{H}_\varepsilon(E) \le \sum_j \mathcal{H}_\varepsilon(E_j)$$

.

PROOF. Let $\{C_{jl}\}$ $l = 1, \dots N_j$ be an ε-covering of E_j. Then the family

$$\{C_{1l_1} \times C_{2l_2} \times \cdots \times C_{kl_k} : l_j = 1, \dots, N_j\}$$

is an ε-covering of E. □

We also need upper bounds for ε-entropy of a ball in finite-dimensional ℓ^∞ space. Let $\mathbb{R}^n_\infty$ be $\mathbb{R}^n$ with the sup-norm:

$$||(x_1, x_2, \dots, x_n)||_\infty = \max_j |x_j| \, .$$

LEMMA 3.3. *Let B_r be a ball of radius r in $\mathbb{R}^n_\infty$. Then*

$$\mathcal{H}_\varepsilon(B_r) < n \log\left(\frac{r}{\varepsilon} + 1\right) \, .$$

PROOF. The inequality is obvious for $n = 1$. The general case then follows from Lemma (3.2). □

4. Kolmogorov Dimension

Let X be a compact subset of a domain $D \subset \mathbb{C}^n$. Let A_X^D be a set of traces on X of functions analytic in D and bounded by 1. So $f \in A_X^D$ if and only if there exists a function F holomorphic on D such that

$$\sup_{z \in D} |F(z)| \le 1$$

and $f(z) = F(z)$ for every $z \in X$. By Montel's theorem A_X^D is a compact subset of $C(X)$.

The connections between the asymptotics of ε-entropy and the pluripotential theory were predicted by Kolmogorov, who conjectured that in the one dimensional case

$$\lim_{\varepsilon \to 0} \frac{\mathcal{H}_\varepsilon(A_X^D)}{\log^2(1/\varepsilon)} = \frac{C(X,D)}{(2\pi)} \, ,$$

where $C(X,D)$ is the condenser capacity. This conjecture was proved simultaneously by K. I. Babenko [**2**] and V. D. Erokhin [**8**] for simply-connected domain D and connected compact X (see also [**9**]). For more general pairs (X,D) the conjecture was proved by Widom [**19**] (simplified proof can be found in [**10**]).

In the multidimensional case Kolmogorov asked to prove the existence of the limit

$$\lim_{\varepsilon \to 0} \frac{\mathcal{H}_\varepsilon(A_X^D)}{\log^{n+1}(1/\varepsilon)}$$

and to calculate it explicitly. V. P. Zahariuta [**20**] showed how the solution of Kolmogorov problem will follow from the existence of the uniform approximation of the relative extremal plurisubharmonic function $u^*_{X,D}$ by multipole pluricomplex Green functions with logarithmic poles in X (Zahariuta conjecture). Later this conjecture was proved by Nivoche [**15**] for a "nice" pairs (D,X). Therefore it is established that for such pairs

$$\lim_{\varepsilon \to 0} \frac{\mathcal{H}_\varepsilon(A_X^D)}{\log^{n+1}(1/\varepsilon)} = \frac{C(X,D)}{(2\pi)^n} \, ,$$

where $C(X,D)$ is the relative capacity (see [**5**]).

The pluripolarity of X is equivalent to the condition $C(X,D) = 0$ ([**5**]). If $\mathcal{H}_\varepsilon(A_X^D) = o(\log^{n+1}(\frac{1}{\varepsilon}))$ then X is "small" (pluripolar) and the asymptotics of an ε-entropy can be used to determine how "small" X is. We will use the function

$$\Psi(X,D) = \limsup_{\varepsilon\to 0} \frac{\log \mathcal{H}_\varepsilon(A_X^D)}{\log\log\frac{1}{\varepsilon}} - 1 \tag{4.1}$$

to characterize the "dimension" of X.

For a compact subset $X \subset \mathbb{C}^n$ we define the *Kolmogorov dimension* $\mathcal{K}$-dim $X = \Psi(X,D)$, where D is a bounded domain containing X. A. N. Kolmogorov proposed in [**12**] to use $\Psi(X,D)$ as a *functional dimension* of space of holomorphic functions on X. The idea to use $\Psi(X,D)$ to characterize the "size" of compact X seems to be new. We proceed to prove that $\Psi(X,D)$ is independent of the bounded domain containing X.

LEMMA 4.1. *Let $D \subset \mathbb{C}^n$ be a bounded domain. If $X_1, X_2, \dots, X_k$ are compact subsets of D, then*

$$\Psi(\bigcup X_j, D) = \max \Psi(X_j, D)\,.$$

PROOF. Let $X = \bigcup X_j$. Embeddings $X_j \to X$ generate a natural isometric embedding $C(X) \to C(X_1) \times C(X_2) \times \dots \times C(X_k)$. Restriction of this embedding on A_X^D gives an isometric embedding $A_X^D \to A_{X_1}^D \times A_{X_2}^D \times \dots \times A_{X_k}^D$. By Lemma 3.2

$$\mathcal{H}_\varepsilon(A_X^D) \le \sum_{j=1}^k \mathcal{H}_\varepsilon(A_{X_j}^D) \le k \max \mathcal{H}_\varepsilon(A_{X_j}^D)\,, \tag{4.2}$$

and the result follows. □

For a point $a \in \mathbb{C}^n$ and $R > 0$, we denote $\Delta(a,R)$ the open polydisk of radius R with center at a

$$\Delta(a,R) = \{z = (z_1, z_2, \dots, z_n) : |z_j - a_j| < R,\ \ j = 1, \dots, n\}\,.$$

The following well known result follows directly from Cauchy's formula.

LEMMA 4.2. *Let R and r be real numbers, $R > r > 0$. Let $a \in \mathbb{C}^n$ and f be a bounded analytic function on the polydisk $\Delta(a,R)$. Then for any positive integer k there exists a polynomial P_k of degree k such that*

$$\sup_{w\in\Delta(a,r)} |f(w) - P_k(w)| \le \frac{1}{R-r}\left(\frac{r}{R}\right)^k \sup_{z\in\Delta(a,R)} |f(z)|\,. \tag{4.3}$$

LEMMA 4.3. *Let R and r be real numbers, $R > r > 0$. If a polynomial P of degree k satisfies the inequality $|P(w)| \le A$ for every $w \in \Delta(a,r)$, then for $z \in \Delta(a,R)$ we have*

$$|P(z)| \le A\left(\frac{R}{r}\right)^k\,. \tag{4.4}$$

PROOF. Let $Q(\lambda) = P((\lambda(z-a)+a)$. The inequality follows from the application of maximum modulus principle to $Q(\lambda)/\lambda^k$. □

THEOREM 4.4. *Let X be a compact in $\mathbb{C}^n$. If D_1, $D_2 \subset \mathbb{C}^n$ are bounded domains containing X, then $\Psi(X,D_1) = \Psi(X,D_2)$.*

PROOF. Without loss of generality we may assume that $D_1 \subset D_2$. Then $A_X^{D_1} \supset A_X^{D_2}$ and $\Psi(X, D_1) \geq \Psi(X, D_2)$. We establish the special case of two polydisks first. Suppose $D_1 = \Delta(a, r)$ and $D_2 = \Delta(a, R)$, where $R > r > 0$. Choose $r' > 0$ such that $r > r'$ and $X \subset \Delta(a, r')$. Let $\{f_1, f_2, \dots, f_N\} \subset A_X^{D_1}$ be a maximal ε-distinguishable set with $N = M_\varepsilon(A_X^{D_1})$. By Lemma 4.2 there exist a positive integer k and polynomials $\{p_1, p_2, \dots, p_N\}$ of degree k such that

$$\sup_{z \in X} |f_j(z) - p_j(z)| < \varepsilon/3 \qquad \text{for } j = 1, 2, \dots, N$$

and $k \leq L \log \frac{1}{\varepsilon}$, where L depends only on R and r'. Then polynomials $\{p_1, p_2, \dots, p_N\}$ are $\varepsilon/3$-distinguishable. By Lemma 4.3, polynomials

$$q_j = \left(\frac{r}{R}\right)^k p_j$$

are bounded on D_2 by 1. There exist positive constants c, λ, which depend only on R, r and r', but not on ε, such that polynomials $\{q_1, q_2, \dots, q_N\}$ are δ-distinguishable (as points in $A_X^{D_2}$), where $\delta = c\varepsilon^\lambda$. Hence

$$N_\delta(A_X^{D_2}) \leq M_\delta(A_X^{D_2}) \leq N_\varepsilon(A_X^{D_1}) \,, \tag{4.5}$$

and $\Psi(X, D_1) = \Psi(X, D_2)$.

If $D_1 = \Delta(a, r)$ and D_2 is an arbitrary bounded domain containing D_1, then there exists $R > 0$, such that $D_3 = \Delta(a, R) \supset D_2$. In this case the theorem follows from the inequalities

$$\Psi(X, D_1) \geq \Psi(X, D_2) \geq \Psi(X, D_3) = \Psi(X, D_1) \,.$$

Now consider the general case. Let polydisks $\Delta_1, \Delta_2, \dots, \Delta_s \subset D_1$ form an open cover of X. There exist compact sets $X_1, X_2, \dots X_s$ such that $X_j \subset \Delta_j$ for $j = 1, 2, \dots, s$ and $\bigcup X_j = X$. Then

$$\Psi(X_j, D_1) = \Psi(X_j, \Delta_j) = \Psi(X_j, D_2) \,, \qquad \text{for } j = 1, 2, \dots, s.$$

The theorem follows now from Lemma 4.1. □

EXAMPLE 4.5. Let $X = \overline{\Delta(0, r)}$. Let $R > r$ and $D = \Delta(0, R)$. Kolmogorov [**12**] (see also [**14**]) showed that

$$\mathcal{H}_\varepsilon(A_X^D) = C(n, r, R) \left(\log \frac{1}{\varepsilon}\right)^{n+1} + \mathrm{O}\left(\left(\log \frac{1}{\varepsilon}\right)^n \log\log \frac{1}{\varepsilon}\right) \,. \tag{4.6}$$

Therefore $\Psi(X, D) = n$ and $\mathcal{K}$-dim $X = n$.

THEOREM 4.6. *Let X be a compact subset of $\mathbb{C}^n$. The Kolmogorov dimension $\mathcal{K}$-dim X satisfies the following properties.*

1. $0 \leq \mathcal{K}$*-dim* $X \leq n$.
2. $\mathcal{K}$*-dim* $\{z\} = 0$.
3. $\mathcal{K}$*-dim* $\widehat{X} = \mathcal{K}$*-dim* X.
4. *If* $Y \subset X$ *then* $\mathcal{K}$*-dim* $Y \leq \mathcal{K}$*-dim* X.
5. *If* $\{X_j\}$ *is a finite family of compact subsets of* $\mathbb{C}^n$, *then*
$$\mathcal{K}\text{-}dim \bigcup_{j=1}^m X_j = \max\{\mathcal{K}\text{-}dim\ X_j : j = 1, \dots, m\} \,.$$
6. *If* $D \subset \mathbb{C}^n$ *is a domain,* $X \subset D$ *and* $\phi : D \to \mathbb{C}^k$ *is a holomorphic map, then* $\mathcal{K}$*-dim* $\phi(X) \leq \mathcal{K}$*-dim* X.

(7) *If $\mathcal{K}$-dim $X < n$, then X is pluripolar.*

REMARK 4.7. Property (5) does not hold for countable unions. There exists a countable compact set X such that $n = \mathcal{K}$-dim X (see Example 4.10). Such set X also provides a counterexample to the converse of (7).

PROOF. Properties (4) and (6) follow immediately from the definition. Property (2) follows from Lemma 3.3. Lemma 4.1 implies (5). The inequality $\mathcal{K}$-dim $X \geq 0$ immediately follows from the definition. From (4.6) follows that Kolmogorov dimension of a closed polydisk equal n. Therefore (4) implies the second part of (1).

To show (3) consider a Runge domain D containing X. Let $W = \widehat{X}$. Then A_X^D and A_W^D are isometric, hence $\Psi(X, D) = \Psi(\widehat{X}, D)$ and (3) follows.

The property (7) follows from the following theorem and Lemma 4.11. □

THEOREM 4.8. *Let X be a compact subset of $\mathbb{C}^n$, such that $\mathcal{K}$-dim $X = s < n$. Then for every $1 < h < \frac{n}{s}$ and every $N > N_0 = N_0(h)$ there exists a non-constant polynomial $P \in \mathbb{Z}[z_1, z_2, \ldots, z_n]$, $\deg P \leq N$ with coefficients bounded by $\exp(N^h)$, such that*

$$\sup\{|P(z)| : z \in X\} < \exp(-N^h)\ . \tag{4.7}$$

PROOF. The result follows from Dirichlet principle. Let $D = \Delta(0, R)$ be a polydisk containing X. Assume that $R > 1$. Let $\varepsilon = \frac{1}{2} exp(-2N^h - N\log R - n\log N)$. Choose t such that $s < t < \frac{n}{h}$. For large enough N there exists an ε-covering of A_X^D with cardinality $\leq \exp\left\{(\log\frac{1}{\varepsilon})^{t+1}\right\}$.

Let $T = [exp(n^h)]$ (integer part of $exp(n^h)$). There are

$$M = T^{\binom{N+n}{n}}$$

polynomials of degree at most N with coefficients in $\{1, 2, \ldots T\}$. Let $\{p_1, p_2, \ldots p_M\}$ be a list of all such polynomials. Clearly the polynomial

$$q_j = \frac{1}{N^n R^N \exp(N^h)} p_j$$

belongs to A_X^D. By our choice of t, $h(t+1) < n + h$, therefore there are more polynomials q_j than cardinality of the ε-covering, so there are two polynomials, let say q_1 and q_2 such that

$$|q_1(z) - q_2(z)| \leq 2\varepsilon \qquad \text{for every } z \in X.$$

Then $P = p_1 - p_2$ satisfies (4.7). □

Let $\mathcal{P}_N$ be the set of all polynomials (with complex coefficients) on $\mathbb{C}^n$ of the degree $\leq N$, whose supremum on a unit polydisk $\Delta(0, 1)$ is at least 1.

COROLLARY 4.9. *If $\mathcal{K}$-dim $X = s < n$, then for every $1 < h < \frac{n}{s}$ and every $N > N_0 = N_0(h)$ there exists polynomial $P \in \mathcal{P}_N$, such that*

$$\sup\{|P(z)| : z \in X\} < \exp(-N^h)\ . \tag{4.8}$$

Corollary 4.9 may be used to bound Kolmogorov dimension from below.

EXAMPLE 4.10. Given $0 < r < 1$ and a positive integer N there exists a finite set $X_{r,N} \subset \Delta(0,r)$, such that for any polynomial $P \in \mathcal{P}_N$, the following inequality holds

$$\max_{z \in X_{r,N}} |P(z)| \geq \frac{1}{2} r^N . \tag{4.9}$$

For example, if $\varepsilon = \frac{1-r}{2N} r^N$, then a maximal ε-distinguishable subset of $\Delta(0,r)$ satisfies condition (4.9). Let

$$X = \bigcup_{k=2}^{\infty} X_{1/k,k} .$$

Then X is compact and for any $N > 2$ and $P \in \mathcal{P}_N$

$$\sup_{z \in X} |P(z)| \geq \max_{z \in X_{1/N,N}} |P(z)| \geq \frac{1}{2} \left(\frac{1}{N} \right)^N .$$

Therefore by Corollary 4.9 $\mathcal{K}$-dim $X = n$.

To finish the proof of Theorem 4.6 we need the following well-known result.

LEMMA 4.11. *Let X be a compact subset of $\mathbb{C}^n$. If there exists a sequence $\{a_k\}$, $a_k > 0$ and a family of polynomials $P_k \in \mathcal{P}_k$ such that*

$$\sup_{z \in X} |P_k(z)| \leq e^{-a_k} , \qquad \text{and} \tag{4.10}$$

$$\lim \frac{a_k}{k} = \infty , \tag{4.11}$$

then X is pluripolar.

PROOF. Let $v_k(z) = \frac{1}{a_k} \log P_k(z)$ and $v(z) = \limsup v_k(z)$. We will show that $v \geq -2/3$ on a dense set. Let $\zeta \in \mathbb{C}^n$ and $0 < \delta < 1$. Suppose that $\Delta(\zeta, R) \supset \Delta(0,1)$. We will show that there exists a nested sequence of closed polydisks $\Delta_m = \overline{\Delta(w_m, \delta_m)}$ with $\Delta_1 = \overline{\Delta(\zeta, \delta)}$, and an increasing sequence of positive integers $k_1 = 1 < k_2 < \cdots < k_m < \ldots$ such that $v_{k_m} \geq -2/3$ on Δ_m for $m > 1$. Given $\Delta = \Delta_m = \overline{\Delta(w_m, \delta_m)}$ by Lemma 4.3 for any given k there exists $w \in \Delta$ such that

$$|P_k(w)| \geq \left(\frac{\delta_m}{R + \delta} \right)^k . \tag{4.12}$$

Choose $k = k_{m+1} > k_m$ such that

$$\frac{a_k}{k} \geq 2 \log \frac{R + \delta}{\delta_m} .$$

Then by (4.12) $v_k(w) \geq -1/2$. Choose $w_{m+1} = w$. Because the function v_k is continuous at w, there exists a closed polydisk $\Delta_{m+1} = \overline{\Delta(w_{m+1}, \delta_{m+1})} \subset \Delta_m$, such that $v_k \geq -2/3$ on Δ_{m+1}. Therefore $v \geq -2/3$ on a dense set. By (4.10), $v|_X \leq -1$ and so X is a negligible set. By [**5**], negligible sets are pluripolar and result follows. ☐

REMARK 4.12. This lemma and the converse follow from Theorem 2.1 in [**1**].

5. Manifolds of Gevrey Class

In view of Theorem 4.8 and Theorem 4.6 (6), Theorems 1.2 and 1.1 are corollaries of Theorem 1.3. In this section we prove Theorem 1.3.

Let $M \subset \mathbb{C}^n$ be an m-dimensional totally real submanifold of Gevrey class G^s. Let $X \subset M$ be a compact subset. Fix $p \in M$. There exist holomorphic coordinates $(z, w) = (x + iy, w) \in \mathbb{C}^n$, $x,y \in \mathbb{R}^m$, $w \in \mathbb{C}^{n-m}$ near p, vanishing at p, real-valued functions of class G^s h_1, $h_2, \dots, h_m$, and complex valued functions of class G^s H_1, $H_2, \dots, H_{n-m}$ such that $h_1'(0) = h_2'(0) = \dots = h_m'(0) = 0$, $H_1'(0) = H_2'(0) = \dots = H_{n-m}'(0) = 0$, and locally

$$M = \{(x + iy, w) : y_j = h_j(x), w_k = H_k(x)\} \ . \tag{5.1}$$

For smooth manifold the existence of such coordinates is well known (see, for example [**3**], Proposition 1.3.8). Note, that functions h_j and H_k are defined by Implicit Function Theorem, and so by [**13**] are of class G^s.

We fix such coordinates and choose r sufficiently small. In view of Theorem 4.6 (5), it is sufficient to prove Theorem 1.3 for $X \subset \Delta(p, r)$. Put $D = \Delta(p, 1)$. To estimate $\Psi(X, D)$ we will cover X by small balls, approximate functions in A_X^D by Taylor polynomials, and then replace in these polynomials terms w^λ and y^ν by Taylor polynomials of functions H^λ and h^ν. To estimate the Taylor coefficients for powers of functions of Gevrey class we need the following lemma.

LEMMA 5.1. *If $f \in G^s(K)$ and $|f| \le 1$ on K, then there exist a constant C such that for any positive integer k and any multi-index α the following inequality holds on K*

$$|\partial^\alpha f^k| \le C^{|\alpha|} \binom{\alpha + k - 1}{\alpha} (\alpha!)^s \ . \tag{5.2}$$

Recall, that $\alpha + k = (\alpha_1 + k, \alpha_2 + k, \dots, \alpha_m + k)$.

PROOF. We will argue by induction on k. Because $|f| \le 1$, there exists a constant C, such that

$$\partial^\alpha f \le C^{|\alpha|} (\alpha!)^s$$

and (5.2) holds for $k = 1$. Suppose (5.2) holds for $1, 2, \dots, k$, then

$$\begin{aligned} |\partial^\alpha f^{k+1}| &= \left| \sum_{\nu \le \alpha} \binom{\alpha}{\nu} \partial^\nu f^k \partial^{\alpha - \nu} f \right| \le C^{|\alpha|} \sum_{\nu \le \alpha} \binom{\alpha}{\nu} \binom{\nu + k - 1}{\nu} (\nu!)^s \left((\alpha - \nu)!\right)^s \\ &= C^{|\alpha|} \sum_{\nu \le \alpha} \binom{\nu + k - 1}{\nu} \binom{\alpha}{\nu} \nu!(\alpha - \nu)!(\nu!)^{s-1} \left((\alpha - \nu)!\right)^{s-1} \\ &\le C^{|\alpha|} \sum_{\nu \le \alpha} \binom{\nu + k - 1}{\nu} \alpha!(\alpha!)^{s-1} = C^{|\alpha|} \binom{\alpha + k}{\alpha} (\alpha!)^s \end{aligned}$$

□

REMARK 5.2. The same proof holds for the product of k different functions, provided that they satisfy the Gevrey class condition (2.1) with the same constant C_K.

Let $t > s \ge 1$ and N be a large integer, which will tend to infinity later. Fix positive $a < t - s$. Put $\delta = N^{1-t}$ and $\varepsilon = N^{-aN}$. We may cover X by less than

$(1/\delta)^m$ balls of radius δ. Let Q be one of these balls and K be the set of restrictions on Q of functions in A^D_X. We claim that any function f in K may be approximated by polynomials in $x_1, x_2, \ldots, x_m$ of the degree $\leq N$ with coefficients bounded by $C^N(N!)^{s-1}$ with error less than 2ε, where constant C depends on X and r only. Let us show how the theorem follows from this claim. The real dimension of the space of polynomials of the degree $\leq N$ is $T = 2\binom{N+m}{N}$. Consider in the T-dimensional space with the sup-norm $\mathbb{R}^T_\infty$ the ball B of a radius $C^N(N!)^{s-1}$. By Lemma 3.3,

$$\mathcal{H}_\varepsilon(B) \leq 2\binom{N+m}{N} \log\left(\frac{C^N(N!)^{s-1}}{\varepsilon} + 1\right) = O(N^{m+1}\log N) .$$

By the claim ε-covering of B generate 3ε-covering of K, therefore

$$\mathcal{H}_{3\varepsilon}(K) = O(N^{m+1}\log N) .$$

Then by (4.2)

$$\mathcal{H}_\varepsilon(A^D_X) = O\left(\left(\frac{1}{\delta}\right)^m N^{m+1}\log N\right) = O(N^{mt+1}\log N) . \tag{5.3}$$

Now we let N tend to infinity. By (5.3), $\mathcal{K}$-dim $X = \Psi(X, D) \leq mt$. The only restriction imposed on t so far was $t > s$. Hence $\mathcal{K}$-dim $X \leq ms$. It remains to prove the claim. We approximate a function f in K in two steps. Consider the Taylor polynomial P of f centered at the center of the ball Q of the degree N . By Cauchy's formula

$$\sup_{Q_l} |f - P| < \frac{1}{1-r-\delta}\left(\frac{\delta}{1-r}\right)^N < \varepsilon$$

for sufficiently large N. Suppose $P(z, w) = \sum c_{\lambda\mu\nu} x^\lambda y^\mu w^\nu$. Because $f \in A^D_X$, $|c_{\lambda\mu\nu}| \leq 1$.

On the next step we approximate y^μ and w^ν by the Taylor polynomials of the degree N of h^μ and H^ν. Let (x_0, y_0, w_0) be the center of the ball Q. Let g be one of the functions h_1, h_2,... h_m, H_1, H_2,...,H_{n-m} and $L \leq N$. Then by Taylor formula

$$g^L(x_0 + h) = \sum_{|\alpha| \leq N} \partial^\alpha f(x_0)\frac{h^\alpha}{\alpha!} + R_N(x, h) .$$

By Lemma 5.1 for $||h||_\infty < \delta$

$$|R_N(x, h)| \leq C^{N+1}\delta^N \sum_{|\alpha|=N+1} \binom{\alpha + N - 1}{\alpha} (\alpha!)^{s-1} .$$

Therefore $\log |R_N(x, h)| = (s - t + o(1))N \log N$ and claim follows.

References

[1] H. J. Alexander and B. A. Taylor, *Comparison of two capacities in* $\mathbf{C}^n$, Math. Z. **186** (1984), no. 3, 407–417.

[2] K. I. Babenko, *On the entropy of a class of analytic functions*, Nauchn. Dokl. Vyssh. Shkol. Ser. Fiz.-Mat. Nauk (1958), no. 2, 9–16.

[3] M. S. Baouendi, P. Ebenfelt, and L. P. Rothschild, *Real submanifolds in complex space and their mappings*, Princeton Mathematical Series, vol. 47, Princeton University Press, Princeton, NJ, 1999.

[4] E. Bedford, *The operator* $(dd^c)^n$ *on complex spaces*, Seminar Pierre Lelong-Henri Skoda (Analysis), 1980/1981, and Colloquium at Wimereux, May 1981, Lecture Notes in Math., vol. 919, Springer, Berlin, 1982, pp. 294–323.

[5] E. Bedford and B. A. Taylor, *A new capacity for plurisubharmonic functions*, Acta Math. **149** (1982), no. 1-2, 1–40.

[6] D. Coman, N. Levenberg, and E. A. Poletsky, *Quasianalyticity and pluripolarity*, J. Amer. Math. Soc. **18** (2005), no. 2, 239–252.

[7] K. Diederich and J. E. Fornæss, *A smooth curve in* $\mathbf{C}^2$ *which is not a pluripolar set*, Duke Math. J. **49** (1982), no. 4, 931–936.

[8] V. D. Erohin, *Asymptotic theory of the* ε*-entropy of analytic functions*, Dokl. Akad. Nauk SSSR **120** (1958), 949–952.

[9] ———, *Best linear approximation of functions analytically continuable from a given continuum to a given region*, Uspehi Mat. Nauk **23** (1968), no. 1 (139), 91–132.

[10] S. D. Fisher and C. A. Micchelli, *The* n*-width of sets of analytic functions*, Duke Math. J. **47** (1980), no. 4, 789–801.

[11] L. Hörmander, *An introduction to complex analysis in several variables*, third ed., North-Holland Publishing Co., Amsterdam, 1990.

[12] A. N. Kolmogorov and V. M. Tihomirov, ε*-entropy and* ε*-capacity of sets in functional space*, Amer. Math. Soc. Transl. (2) **17** (1961), 277–364.

[13] H. Komatsu, *The implicit function theorem for ultradifferentiable mappings*, Proc. Japan Acad. Ser. A Math. Sci. **55** (1979), no. 3, 69–72.

[14] G. G. Lorentz, M. v. Golitschek, and Y. Makovoz, *Constructive approximation*, Springer-Verlag, Berlin, 1996, Advanced problems.

[15] S. Nivoche, *Proof of a conjecture of Zahariuta concerning a problem of Kolmogorov on the* ϵ*-entropy*, Invent. Math. **158** (2004), no. 2, 413–450.

[16] S. I. Pinčuk, *A boundary uniqueness theorem for holomorphic functions of several complex variables*, Mat. Zametki **15** (1974), 205–212.

[17] A. Sadullaev, *A boundary uniqueness theorem in* $\mathbf{C}^n$, Mat. Sb. (N.S.) **101(143)** (1976), no. 4, 568–583, 639.

[18] M. Waldschmidt, *Diophantine approximation on linear algebraic groups*, Springer-Verlag, Berlin, 2000.

[19] H. Widom, *Rational approximation and* n*-dimensional diameter*, J. Approximation Theory **5** (1972), 343–361.

[20] V. P. Zahariuta, *Spaces of analytic functions and maximal plurisubharmonic functions*, Doc. Sci. Thesis, Rostov-on-Don, 1984.

Department of Mathematics and Statistics, University Of New Hampshire, Durham, New Hampshire 03824

E-mail address: oleg.eroshkin@unh.edu

Contemporary Mathematics
Volume **454**, 2008

On a question of Brézis and Korevaar concerning a class of square-summable sequences

Richard Fournier and Luis Salinas

Abstract. We give an new proof of a result due to Brézis and Nirenberg: $\sum_{k=-\infty}^{\infty} k|a_k|^2$ is an integer whenever $\{a_k\}_{k=-\infty}^{\infty}$ is a sequence of complex numbers such that $\sum_{k=-\infty}^{\infty} a_k\bar{a}_{n+k} = \begin{cases} 0 & \text{if } n \neq 0, \\ 1 & \text{if } n = 0, \end{cases}$ for all integers n and $\sum_{k=-\infty}^{\infty} |k|\,|a_k|^2 < \infty$.

1. Introduction

We consider sequences $\{a_n\}_{-\infty}^{\infty}$ of complex numbers such that

$$\sum_{k=-\infty}^{\infty} a_k\bar{a}_{n+k} = \begin{cases} 0 & \text{if } n \neq 0, \\ 1 & \text{if } n = 0, \end{cases} \quad \text{for all integers } n, \tag{1}$$

and

$$\sum_{k=-\infty}^{\infty} |k|\,|a_k|^2 < \infty. \tag{2}$$

Under these assumptions it has been proved by Brézis and Nirenberg [**3, 4**] that the sum of the series $\sum_{k=-\infty}^{\infty} k|a_k|^2$ is an integer, a rather unexpected and sparkling result. The motivation of Brézis and Nirenberg while proving this was to extend the notion of degree (i.e., index or winding number) to various classes of maps; their proof was rather indirect and used aspects of duality. In a remarkable paper [**8**] Korevaar studied what happens to the Brézis–Nirenberg result when the absolute convergence of the series in (2) is replaced by various notions of convergence of $\sum_{k=-\infty}^{\infty} k|a_k|^2$. In the same paper, Korevaar asked for a more direct proof of the result and the same question has been recently raised by Brézis during a talk at a meeting (2004) held in honour of Prof. Andrzej Granas on the occasion of his 75th

---

2000 *Mathematics Subject Classification.* Primary: 42A16; Secondary: 30B10, 30A78.

Key words and phrases. Fourier coefficients of unimodular functions, *Hp* spaces, Sobolev spaces.

R. Fournier was supported by NSERC and L. Salinas by FONDECYT. Both authors would like to thank Oliver Roth and St. Ruscheweyh for their involvement in this project.

birthday. Even more recently, the very same question has been raised by Brézis in [**2**].

It is of course not so clear what is meant by a more direct proof. Our goal in this paper is to provide a different proof of the result based on facts more readily evident to "classical" complex analysts. Our work is also related to remarks of L. Boutet de Monvel and O. Gabber to be found in an appendix to the paper [**1**]. We shall finally also obtain the following

THEOREM 1.1. *Let $\{a_k\}_0^\infty$ be a sequence of complex numbers for which*

$$\sum_{k=0}^{\infty} a_k \bar{a}_{n+k} = \begin{cases} 0 & \text{if } n \neq 0, \\ 1 & \text{if } n = 0, \end{cases} \quad \text{for all positive integers } n \text{ and } \sum_{k=0}^{\infty} |k|\, |a_k|^2 < \infty.$$

Then $B(z) := \sum_{k=0}^{\infty} a_k z^k$ is a finite Blaschke product and the number of zeros of B in the unit disc $\{z \mid |z| < 1\}$, including multiplicities, is equal to $\sum_{k=1}^{\infty} k|a_k|^2$.

2. Another Proof of the Brézis–Nirenberg Result

We shall proceed by a number of lemmas.

LEMMA 2.1. *Under the hypothesis* (1), *the 2π-periodic function $u(\theta) := \sum_{k=-\infty}^{\infty} a_k e^{ik\theta}$ is well-defined and unimodular for almost all $\theta \in [0, 2\pi)$.*

PROOF. By the Riesz–Fischer theorem, there exists an integrable function u whose Fourier coefficients are the numbers $\{a_n\}_{n=-\infty}^{\infty}$ and by the famous result of Carleson, this function is almost everywhere equal to its Fourier series. (This may also be established by using an older and weaker result of Fejér [**10**, p. 65]). Thus, we may assume that $u(\theta) := \sum_{k=-\infty}^{\infty} a_k e^{ik\theta}$ is the Fourier series of a square summable function. We now define, for $0 < r < 1$,

$$u_r(\theta) = \sum_{k=-\infty}^{\infty} a_k r^{|k|} e^{ik\theta}, \quad 0 \leq \theta < 2\pi.$$

This last series is absolutely and uniformly convergent. We have for each integer n,

$$\begin{aligned}
&\left| \frac{1}{2\pi} \int_0^{2\pi} e^{in\theta} |u_r(\theta)|^2 \, d\theta - \frac{1}{2\pi} \int_0^{2\pi} e^{in\theta} |u(\theta)|^2 \, d\theta \right| \\
&\leq \frac{1}{2\pi} \int_0^{2\pi} |u_r(\theta)^2 - u(\theta)^2| \, d\theta \leq \frac{1}{2\pi} \int_0^{2\pi} (|u_r(\theta)| + |u(\theta)|)\, |u_r(\theta) - u(\theta)| \, d\theta \\
&\leq \left(\left(\frac{1}{2\pi} \int_0^{2\pi} |u_r(\theta)|^2 d\theta \right)^{1/2} + \left(\frac{1}{2\pi} \int_0^{2\pi} |u(\theta)|^2 d\theta \right)^{1/2} \right) \left(\frac{1}{2\pi} \int_0^{2\pi} |u_r(\theta) - u(\theta)|^2 d\theta \right)^{1/2} \\
&\leq 2 \left(\sum_{k=-\infty}^{\infty} |a_k|^2 \right)^{1/2} \left(\sum_{k=-\infty}^{\infty} |a_k|^2 (1 - r^{2|k|}) \right)^{1/2}.
\end{aligned}$$

Since by Abel's continuity theorem

$$\lim_{r \to 1} \sum_{k=-\infty}^{\infty} |a_k|^2 (1 - r^{2|k|}) = 0,$$

we obtain

$$\frac{1}{2\pi}\int_0^{2\pi} e^{in\theta}|u(\theta)|^2\,d\theta = \lim_{r\to 1}\frac{1}{2\pi}\int_0^{2\pi} e^{in\theta}|u_r(\theta)|^2\,d\theta,$$

$$= \lim_{r\to 1}\frac{1}{2\pi}\int_0^{2\pi} e^{in\theta}\left(\sum_{k=-\infty}^{\infty} a_k r^{|k|}e^{ik\theta}\right)\left(\sum_{k=-\infty}^{\infty} \bar{a}_k r^{|k|}e^{-ik\theta}\right)d\theta \tag{3}$$

$$= \lim_{r\to 1}\sum_{k=-\infty}^{\infty} a_k\bar{a}_{k+n}r^{|k|+|k+n|} \tag{4}$$

$$= \sum_{k=-\infty}^{\infty} a_k\bar{a}_{k+n} = \begin{cases} 0 & \text{if } n\neq 0,\\ 1 & \text{if } n=0,\end{cases} \tag{5}$$

the passage from (3) to (4) being justified by the absolute and uniform convergence of the Fourier series $u_r(\theta)$ while (5) follows again from Abel's continuity theorem. This completes the proof of Lemma 2.1: we have shown that all Fourier coefficients (except for the constant one) of $|u|^2$ are zero and thus $|u|^2$ is constant almost everywhere. This result may not be entirely new since a (weaker) version of it was stated without proof in a 1962 paper by Newman and Shapiro [**12**]. Moreover, the condition (1) is in fact equivalent to the unimodularity of the associated function $u(\theta)$: this is also a consequence of Parseval's identity. We may now write

$$u(\theta) = e^{iU(\theta)}, \text{ with } U(\theta) \text{ real for almost all } \theta\in[0,2\pi). \qquad \square$$

Let $U(\theta) := \sum_{k=-\infty}^{\infty} b_k e^{ik\theta}$. It is readily seen from the formula of Devinatz [**13**, p. 328] that

$$\begin{aligned} 4\pi^2\sum_{k=-\infty}^{\infty}|k|\,|a_k|^2 &= \int_0^{2\pi}\int_0^{2\pi}\left|\frac{u(\theta)-u(\varphi)}{e^{i\theta}-e^{i\varphi}}\right|^2 d\theta\,d\varphi\\ &= 4\int_0^{2\pi}\int_0^{2\pi}\frac{\sin^2((U(\theta)-U(\varphi))/2)}{|e^{i\theta}-e^{i\varphi}|^2}\,d\theta\,d\varphi\\ &\le \int_0^{2\pi}\int_0^{2\pi}\left|\frac{U(\theta)-U(\varphi)}{e^{i\theta}-e^{i\varphi}}\right|^2 d\theta\,d\varphi\\ &= 4\pi^2\sum_{k=-\infty}^{\infty}|k|\,|b_k|^2.\end{aligned}$$

Unfortunately, it is in general false that $\sum_{k=-\infty}^{\infty}|k|\,|b_k|^2<\infty$; an easy counterexample is $\nu(\theta) := e^{i\theta}$ since

$$\int_0^{2\pi}\int_0^{2\pi}\left|\frac{\theta-\varphi}{e^{i\theta}-e^{i\varphi}}\right|^2 d\theta\,d\varphi = \infty.$$

The next result we shall need must be well known (in fact a stronger result can be found in [**1**, pp. 21–22]) and we therefore state it without proof:

LEMMA 2.2. *There exists a sequence* $\{u_N\}_1^\infty$ *of continuously differentiable unimodular functions* $u_N(\theta) := \sum_{k=-\infty}^{\infty} a_k(N)e^{ik\theta}$ *such that*

$$\lim_{N\to\infty}\sum_{k=-\infty}^{\infty}(|k|+1)\,|a_k-a_k(N)|^2 = 0$$

and

$$\sum_{k=-\infty}^{\infty} |k|\,|a_k(N)|^2 < \infty, \quad N \geq 1.$$

Also related to Lemma 2.2 is an older result due to Krein [**9**] stating in particular that the class $H^{1/2}$ (see the definition below) is a Banach algebra closed under composition with certain analytic functions.

We shall now prove that the result of Brézis and Nirenberg holds for sequences $\{a_n\}_{n=-\infty}^{\infty}$ corresponding to smooth enough functions $u(\theta) := \sum_{k=-\infty}^{\infty} a_k e^{ik\theta}$.

LEMMA 2.3. *Let $u(\theta) := \sum_{k=-\infty}^{\infty} a_k e^{ik\theta}$ where $\{a_k\}$ satisfies* (1) *and* (2) *and with a continuous derivative $u'(\theta)$ over $[0, 2\pi]$. Then $\sum_{k=-\infty}^{\infty} k|a_k|^2$ is an integer.*

PROOF. Let $\phi = 1/2\pi i \int_0^{2\pi} \overline{u(t)} u'(t)\, dt$. By using the functions $u_r(\theta)$ as in the proof of Lemma 2.1, we obtain

$$\phi = \sum_{k=-\infty}^{\infty} k|a_k|^2.$$

Let us also define $\lambda(\theta) := \int_0^{\theta} \overline{u(t)} u'(t)\, dt$ and $\varphi(\theta) := u(\theta) e^{-\lambda(\theta)}$. Then by Lemma 2.1

$$\varphi'(\theta) = (1 - |u(\theta)|^2) u'(\theta) e^{-\lambda(\theta)} \equiv 0, \quad \theta \in [0, 2\pi].$$

We have that $\varphi(\theta) \equiv \varphi(0) = u(0) = u(2\pi)$ and $e^{\lambda(2\pi)} = e^{2\pi i \phi} = 1$. The claim follows. □

We may now prove the general Brézis–Nirenberg result. Our arguments become clearer when expressed within the frame of certain Sobolev spaces (although we shall not use any of the deeper results concerning these spaces). Let $H^{1/2}$ denote the set of all 2π–periodic Fourier series $\sum_{k=-\infty}^{\infty} c_k e^{ik\theta}$ which satisfy $\sum_{k=-\infty}^{\infty} |k|\,|c_k|^2 < \infty$. It is known [**11**, pp. 246–247] that $H^{1/2}$ can be turned into a Banach space when endowed with the norm

$$\left\| \sum_{k=-\infty}^{\infty} c_k e^{ik\theta} \right\| := \left(\sum_{k=-\infty}^{\infty} (|k|+1)|c_k|^2 \right)^{1/2}.$$

For example Lemma 2.2 amounts to the fact that the subspace of smooth functions in $H^{1/2}$ is dense in $H^{1/2}$. Let us consider a sequence $\{a_n\}_{n=-\infty}^{\infty}$ satisfying (1), (2) and $u(\theta) := \sum_{n=-\infty}^{\infty} a_n e^{in\theta}$; let also $\{u_N\}_{N=1}^{\infty}$, $u_N(\theta) := \sum_{k=-\infty}^{\infty} a_k(N) e^{ik\theta}$, be the sequence of smooth functions given by Lemma 2.2. Because $\lim_{N\to\infty} \|u_N - u\| = 0$, we have that $\|u_N\|$ is bounded above and by Lemma 2.3, the sequence of integers $\sum_{k=-\infty}^{\infty} k|a_k(N)|^2$, $N = 1, 2, \ldots$ is therefore also bounded above; we may therefore assume that for N large enough, $\sum_{k=-\infty}^{\infty} k|a_k(N)|^2 := I$ is a fixed integer depending only on u. Further

$$\begin{aligned} \sum_{k=-\infty}^{\infty} k|a_k(N) - a_k|^2 &= -\sum_{k=-\infty}^{\infty} k|a_k(N)|^2 + \sum_{k=-\infty}^{\infty} k|a_k|^2 - 2\sum_{k=-\infty}^{\infty} k\bar{a}_k(N)\big(a_k - a_k(N)\big) \\ &= -I + \sum_{k=-\infty}^{\infty} k|a_k|^2 - 2\sum_{k=-\infty}^{\infty} k\bar{a}_k(N)\big(a_k - a_k(N)\big) \end{aligned}$$

where

$$\left|\sum_{k=-\infty}^{\infty} k|a_k(N) - a_k|^2\right| \leq \|u_N - u\|^2$$

and

$$\begin{aligned}\left|2\sum_{k=-\infty}^{\infty} k\bar{a}_k(N)\big(a_k - a_k(N)\big)\right| &\leq 2\sum_{k=-\infty}^{\infty} |k|\,|a_k(N)|\,|a_k - a_k(N)| \\ &\leq 2\left(\sum_{k=-\infty}^{\infty} |k|\,|a_k(N)|^2\right)^{1/2}\left(\sum_{k=-\infty}^{\infty} |k|\,|a_k - a_k(N)|^2\right)^{1/2} \\ &\leq 2\|u_N\|\,\|u - u_N\| \\ &\leq 2(\|u\| + 1)\|u - u_N\|.\end{aligned}$$

We clearly have

$$\sum_{k=-\infty}^{\infty} k|a_k|^2 = I = \sum_{k=-\infty}^{\infty} k|a_k(N)|^2, \quad \text{for all } N \text{ large enough.}$$

3. Proof of Theorem 1.1

We write $W(z) := \sum_{k=0}^{\infty} a_k z^k$ for $z \in \mathbb{D} := \{z \mid |z| < 1\}$. Because $\sum_{k=0}^{\infty} |a_k|^2 < \infty$, W is well-defined and in fact belongs to the Hardy space $H_2(\mathbb{D})$ (see [**7**] for a standard reference concerning Hardy spaces). By Lemma 2.1, (with $a_k = 0$ for $k < 0$), the radial limits (which are known to exist a.e. $[0, 2\pi)$)

$$W(e^{i\theta}) := \lim_{\substack{r \to 1 \\ 0<r<1}} W(re^{i\theta})$$

satisfy $|W(e^{i\theta})| = 1$ almost everywhere, i.e., W is an inner function. If W is neither a constant nor a finite Blaschke product, it is known that W takes on all values in the unit disc $\mathbb{D}$ infinitely many times, except possibly for an exceptional set of planar measure zero (this is due to Frostman [**5**, p. 35]). In other words,

$$\begin{aligned}\infty &= \lim_{r \to 1} \iint_{x^2+y^2 \leq r^2} |W'(x + iy)|^2 \, dx \, dy \\ &= \lim_{r \to 1} \frac{1}{2} \int_0^{2\pi} re^{i\theta} W'(re^{i\theta}) \overline{W(re^{i\theta})} \, d\theta \\ &= \lim_{r \to 1} \pi \sum_{n=1}^{\infty} n|a_n|^2 r^{2n}.\end{aligned}$$

This of course contradicts the hypotheses (2), again because of the Abel continuity theorem. A more refined argument leading to the same result can be found in [**6**, pp. 60–61]. We may therefore assume that W is a constant of modulus 1 or else a finite Blaschke product,

$$W(z) = \zeta \prod_{j=1}^{J} \frac{z - a_j}{1 - \bar{a}_j z}, \quad |a_j| < 1, \; j = 1, 2, \ldots, J, \quad |\zeta| = 1$$

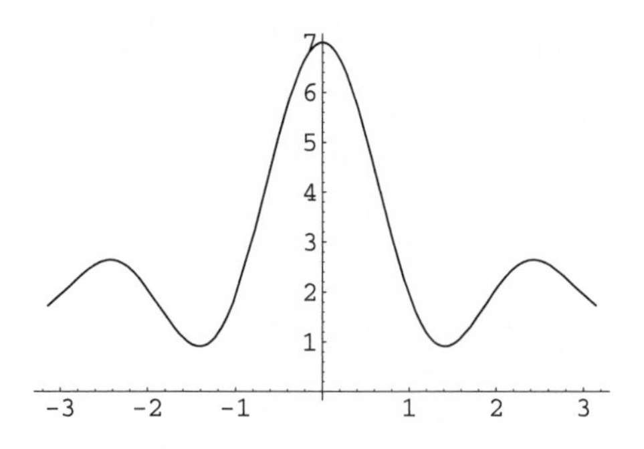

FIGURE 1

and then

$$\begin{aligned}\sum_{n=1}^{\infty} n|a_n|^2 &= \frac{1}{2\pi}\int_0^{2\pi} e^{i\theta}W'(e^{i\theta})\overline{W(e^{i\theta})}\,d\theta \\ &= \frac{1}{2\pi i}\int_{|\zeta|=1} \frac{W'(\zeta)}{W(\zeta)}\,d\zeta \\ &= J,\end{aligned}$$

by the argument principle.

We end this paper by a remark: any unimodular function $u(\theta)$ of the type considered here is a limit of the type

$$u(\theta) = \lim_{j\to\infty} B_{1,j}(e^{i\theta})\overline{B_{2,j}(e^{i\theta})}, \quad \text{a.e. } [-\pi,\pi),$$

where $\{B_{i,j}\}$ and $\{B_{2,j}\}$ are two sequences of Blaschke products. This is a result essentially due to Douglas and Rudin [**7**, p. 153] and one may wonder particularly in view of Theorem 1.1, if

$$u(\theta) = B_1(e^{i\theta})\overline{B_2(e^{i\theta})} \tag{6}$$

where B_1 and B_2 are Blaschke products (or more generally inner functions). We show that this is not true in general; consider for example a real trigonometric polynomial $U_{\mathbb{R}}$ as displayed in Fig. 1 and perturb $U_{\mathbb{R}}(\theta)$ to a real function $U(\theta)$

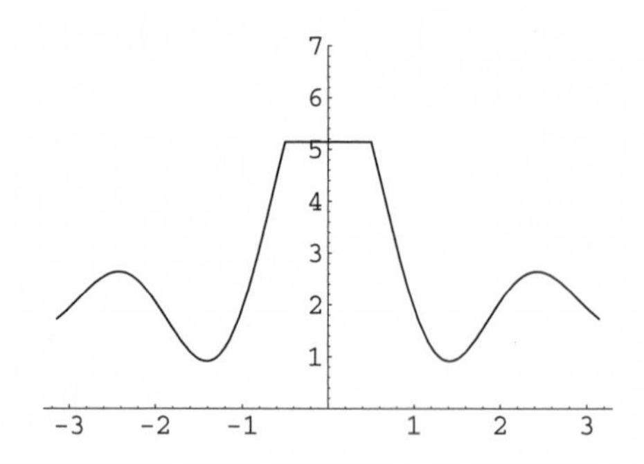

FIGURE 2

whose graph is shown in Fig. 2. Clearly $U_{\mathbb{R}} \in H^{1/2}$ and since

$$|U(\theta) - U(\varphi)| \leq |U_{\mathbb{R}}(\theta) - U_{\mathbb{R}}(\varphi)|, \quad \theta, \varphi \in [-\pi, \pi),$$

we also have $U \in H^{1/2}$. Let $u(\theta) := e^{iU(\theta)}$. By the formula of Devinatz, u belongs to $H^{1/2}$ and if (6) holds for u, the function B_1/B_2 (i.e., a quotient of bounded analytic functions) has constant radial limits almost everywhere on an arc of the unit circle; it is well known [**5**, p. 41] that this can occur only if B_1 is a constant multiple of B_2 and this is ruled out by the definition of U.

References

1. A. Boutet de Monvel–Berthier, V. Georgescu, and R. Purice, *A boundary value problem related to the Ginzburg–Landau model*, Comm. Math. Phys. **142** (1991), no. 1, 1–23.
2. H. Brézis, *New questions related to the topological degree*, The Unity of Mathematics, Progr. Math., vol. 244, Birkhäuser Boston, Boston, MA, 2006, pp. 137–154.
3. H. Brézis and L. Nirenberg, *Degree theory and* BMO. I. *Compact manifolds without boundaries*, Selecta Math. (N.S.) **1** (1995), no. 2, 197–263.
4. ———, *Degree theory and* BMO. II. *Compact manifolds with boundaries*, Selecta Math. (N.S.) **2** (1996), no. 3, 309–368.
5. E. F. Collingwood and A. J. Lohwater, *The theory of cluster sets*, Cambridge Univ. Press, Cambridge, 1966.
6. R.A. Hibschweiler and Thomas H. MacGregor, *Fractional Cauchy Transform*, Chapman and Hall/CRC, USA, 2006.
7. P. Koosis, *Introduction to Hp spaces*, Cambridge Univ. Press, Cambridge, 1998.
8. J. Korevaar, *On a question of Brézis and Nirenberg concerning the degree of circle maps*, Selecta Math. (N.S.) **5** (1999), no. 1, 107–122.
9. M. G. Krein, *On some new Banach algebras and Wiener–Lévy type theorems for Fourier series and integrals*, Amer. Math. Soc. Transl. **93** (1970), 177–199.
10. E. Landau and D. Gaier, *Darstellung und Begründung einiger neuerer Ergebnisse der Funktionentheorie (German) [Presentation and explanation of some more recent results in function theory]*, 3rd ed., Springer-Verlag, Berlin, 1986.
11. R. C. McOwen, *Partial differential equations*, Prentice Hall, Upper Saddle River, 2003.
12. D. J. Newman and H. S. Shapiro, *The Taylor coefficients of inner functions*, Michigan Math. J. **9** (1962), 249–255.
13. B. Simon, *Orthogonal polynomials on the unit circle.* I. *Classical theory*, Amer. Math. Soc. Colloq. Publ., vol 54, Part 1, Amer. Math. Soc., Providence, R.I., 2005.

Département de Mathématiques et Centre de recherches mathématiques, Université de Montréal, C.P. 6128, Succ. Centre-ville, Montréal, Québec H3C 3J7, Canada
E-mail address: fournier@dms.umontreal.ca

Departamento de Informática, Universidad Técnica Federico Santa Maria, Valparaíso, Chile
E-mail address: lsalinas@inf.utfsm.cl

Contemporary Mathematics
Volume **454**, 2008

Approximating $\overline{z}$ in Hardy and Bergman Norms

Zdeňka Guadarrama and Dmitry Khavinson

ABSTRACT. We consider the problem of finding the best analytic approximation in Smirnov and Bergman norm to general monomials of the type $z^n\overline{z}^m$. We show that in the case of approximation to $\overline{z}$ in the annulus (and the disk) the best approximation is the same for all values of p. Moreover, the best approximations to $\overline{z}$ in Smirnov and Bergman spaces characterize disks and annuli.

1. Introduction

Throughout this paper, G denotes a bounded domain in $\mathbb{C}$ with boundary Γ consisting of n simple closed analytic curves. $R(\overline{G})$ will stand for the uniform closure of the algebra of rational functions in G with poles outside of $\overline{G}$.

Let ds be the arclength measure on the boundary of G. Recall that a function f belongs to the Smirnov class $\mathbb{E}_p(G)$ for $1 \leq p < \infty$, if it is analytic in G and there exists a sequence of finitely connected domains $\{G_n\}_{n=1}^{\infty}$, $G_1 \subset G_2 \subset G_3 \subset ...$ with rectifiable boundaries Γ_n so that $\overset{\infty}{\underset{n=1}{\cup}} G_n = G$, and a constant $M > 0$ such that $\|f\|_{\mathbb{E}_p} := \sup\limits_n \left[\int_{\Gamma_n} |f(z)|^p ds\right]^{\frac{1}{p}} \leq M < \infty$. For a nice and concise introduction to Smirnov spaces see [6], also cf. [14].

We let $d\sigma$ denote area measure in G. The Bergman space $\mathbb{A}_p(G)$ for $1 \leq p < \infty$ is the set of analytic functions $f(z)$ in G, with finite norm $\|f\|_{\mathbb{A}_p} = \|f\|_{\mathbb{L}_p(d\sigma, G)} = \left[\int_G |f(z)|^p d\sigma\right]^{\frac{1}{p}}$ (cf. [7]).

D. Khavinson, in [10], [12], [13], [15], [16] posed the question of "how far" $\overline{z}$ is from being approximable by rational functions that are analytic in G. In particular, the following concept was introduced in [15], also cf. [4].

2000 *Mathematics Subject Classification.* 30E10, 41A99.

Key words and phrases. Analytic approximation, Hardy spaces, Bergman spaces.

This work was supported in part by NSF Grant DMS-0139008.

DEFINITION 1.1. *The analytic content* $\lambda(G)$ *of a given domain* G *is defined as*

$$\lambda(G) := \inf_{g\in R(\overline{G})} \|\overline{z} - g(z)\|_{\mathbb{L}_\infty(ds)} = \inf_{g\in R(\overline{G})} \|\overline{z} - g(z)\|_{\mathbb{L}_\infty(d\sigma)}.$$

It turns out that $\lambda(G)$ can be bounded above and below by basic quantities depending on the geometry of the domain G, specifically, its area and perimeter. If we let $A(G)$ denote the area of G and $P(G)$ the perimeter of its boundary, the following inequality holds:

$$\frac{2A\,(G)}{P\,(G)} \leq \lambda(G) \leq \sqrt{\frac{A\,(G)}{\pi}} \tag{1}$$

The upper bound is due to Alexander [3], and the lower bound is due to D. Khavinson [13]. We will refer to this inequality from now on as the A-K inequality.

It follows immediately from (1) that $A\,(G) \leq \frac{P^2(G)}{4\pi}$, which is the isoperimetric inequality. Moreover, when we notice that both inequalities in (1) are sharp, since they become equalities when the domain is a disk, we obtain the isoperimetric theorem (cf. [10]).

The question of what are the extremal domains for the lower bound of (1) still remains open. A few equivalent formulations for the equation $\lambda(G) = \frac{2A(G)}{P(G)}$ in terms of geometry and potential theory can be found in [12] and [15]. The reader may consult the survey [4] which focuses on extremal domains for the left inequality in (1). The following conjecture [4], [14] remains open.

CONJECTURE 1.2. *For a fixed* $\lambda\,(G)$, *the only extremal domains for which the lower bound in* (1) *becomes an equality are the disks of radius* $\lambda\,(G)$, *and annuli* $\{z : r < |z| < R\}$ *with* $\lambda\,(G) = R - r$.

For an extensive discussion about different forms and various ramifications of this conjecture we refer the reader to [4].

If we denote by $\mathbb{E}_1^1(G)$ the unit ball in $\mathbb{E}_1(G)$. We can write

$$\lambda(G) := \inf_{g\in R(\overline{G})} \|\overline{z} - g(z)\|_{\mathbb{L}_\infty(ds,\Gamma)} = \sup_{f\in\mathbb{E}_1^1(G)} \left|\int_\Gamma \overline{z} f(z)dz\right|,$$

and there exist extremal functions $g^*(z)$ and $f^*(z)$ for which the infimum and the supremum above are attained [19]. If the domain is a disk centered at the origin, the best rational approximation to $\overline{z}$ in G is the zero function. In the case of the annulus centered at the origin, the best approximation is $g^*(z) = \frac{Rr}{z}$ [10].

The main focus of this paper is to extend the concept of analytic content to the context of Smirnov and Bergman spaces for $p \geq 1$.

The paper is organized as follows. In the next section we define the Smirnov $p-$analytic content of a domain and show that the A-K inequality extends to the $\mathbb{E}_p$ case yielding bounds for the Smirnov $p-$analytic content in terms of the area and perimeter of the domain. In section 3 we find the best approximation to any monomial $z^n\overline{z}^m$ in the Smirnov $p-$norm of the annulus and the disk. For disks and annuli, the best approximation to $\overline{z}$ turns out to be the same rational function for all p. We prove a converse for this result in the case of the disk, and for $p = 1$ in the case of the annulus. In section 4 we consider the Bergman $p-$analytic content of a domain and explore similar questions, now for the case of the Bergman space $p-$norm. We conclude with some remarks and open questions.

2. Smirnov p-analytic content

DEFINITION 2.1. *The Smirnov $p-$analytic content of a domain G is defined by* $\lambda_{\mathbb{E}_p}(G) := \inf_{g\in\mathbb{E}_p(G)} \|\overline{z} - g(z)\|_{\mathbb{L}_p(ds,\Gamma)}$.

The following general result summarizes the study of extremal problems in Smirnov classes (cf. [19], Theorem 4.3).

COROLLARY 2.2. *Let $p \geq 1$, $\frac{1}{p}+\frac{1}{q} = 1$, and let $\omega(z) \in \mathbb{L}_p(G)$ then the following hold:*
(i) $\inf_{g\in\mathbb{E}_p(G)} \|\omega(z) - g(z)\|_{\mathbb{L}_p(ds,\Gamma)} = \sup_{f\in\mathbb{E}_q^1(G)} \left|\int_\Gamma \omega(z)f(z)dz\right|$.

(ii) There exist extremal functions $g^*(z) \in \mathbb{E}_p(G)$ and $f^*(z) \in \mathbb{E}_q(G)$ for which the infimum and the supremum are attained in (i).
(iii) $g^*(z) \in \mathbb{E}_p(G)$ and $f^*(z) \in \mathbb{E}_q(G)$ are extremal if and only if, almost everywhere on Γ,

$$f^*(z)(\omega(z) - g^*(z))dz = \frac{e^{i\delta}}{\Lambda_{\mathbb{E}_p}^{p-1}}|\omega(z) - g^*(z)|^p ds, \tag{2}$$

where δ is a real constant and $\Lambda_{\mathbb{E}_p} = \|\omega(z) - g^*(z)\|_{\mathbb{L}_p(ds,\Gamma)}$. We will refer to this last equality as the extremality condition in Smirnov spaces.

For $p > 1$ the extremal functions $g^*(z)$ and $f^*(z)$ are unique, the latter up to a factor of $e^{i\alpha}$.

For $p = 1$ the extremal function $f^*(z)$ is unique up to a factor of $e^{i\alpha}$. If the domain G is simply connected, then $g^*(z)$ is also unique. If the domain G is $n-$connected, $n > 1$, then the extremal function $g^*(z)$ is unique provided that $f^*(z)$ has more than $n-2$ zeros in G or that on a certain set $T \subset \Gamma$, $meas(T) > 0$, $|f^*(z)| < 1$. Otherwise, the extremal function $g^*(z)$ may fail to be unique. (cf. Part 3, Theorem 3.2, in [19], and Theorem 3.6 (ii) below).

In our first Theorem we show that following the same strategy used in [10] we can find bounds for $\lambda_{\mathbb{E}_p}(G)$ in terms of the perimeter and the area of the domain G, obtaining the A-K inequality as a limiting case when p approaches infinity.

THEOREM 2.3. *Let $p > 1$, $\frac{1}{p} + \frac{1}{q} = 1$. Let G be a multiply connected domain in $\mathbb{C}$ bounded by n simple closed analytic curves, as before, $A(G)$ denotes the area of G and $P(G)$ the perimeter. Then*

$$\frac{2A(G)}{\sqrt[q]{P(G)}} \leq \lambda_{\mathbb{E}_p}(G) \leq \sqrt{\frac{A\,(G)}{\pi}}\,P(G)^{\frac{1}{p}} \tag{3}$$

If $p = 1$, then $2A(G) \leq \lambda_1(G) \leq \sqrt{\frac{A(G)}{\pi}}P(G)$.

PROOF. We first address the lower bound in (3) for $p > 1$.

$$\lambda_{\mathbb{E}_p}(G) := \|\overline{z} - g^*(z)\|_{\mathbb{L}_p(ds,\Gamma)} = \left(\int_\Gamma |\overline{z} - g^*(z)|^p \frac{ds}{P(G)}\right)^{\frac{1}{p}} (P(G))^{\frac{1}{p}}.$$

By Jensen's inequality, since $p > 1$, we have

$$\begin{aligned}\left(\int_\Gamma |\overline{z} - g^*(z)|^p \frac{ds}{P(G)}\right)^{\frac{1}{p}} (P(G))^{\frac{1}{p}} &\geq \left[\left(\int_\Gamma |\overline{z} - g^*(z)| \frac{ds}{P(G)}\right)^{\frac{1}{p}}\right]^p (P(G))^{\frac{1}{p}} \\ &= P(G)^{-\frac{1}{q}} \int_\Gamma |\overline{z} - g^*(z)| ds \\ &\geq P(G)^{-\frac{1}{q}} \left|\int_\Gamma (\overline{z} - g^*(z)) dz\right|.\end{aligned}$$

Applying the divergence theorem in the form $\int_\Gamma (\overline{z} - g^*(z))dz = 2i\int\int_G \frac{\partial}{\partial \overline{z}}(\overline{z} - g^*(z))d\sigma$, we obtain

$$\lambda_{\mathbb{E}_p}(G) \geq \frac{2A\,(G)}{\sqrt[q]{P(G)}}.$$

For $p = 1$ we have $\lambda_{\mathbb{E}_1}(G) = \int_\Gamma |\overline{z} - g^*(z)| ds \geq \left|\int_\Gamma (\overline{z} - g^*(z))dz\right| = 2A\,(G)$.

Now for the upper bound, and any $p \geq 1$, we will use Corollary 2.2 (i) and by duality rewrite $\lambda_{\mathbb{E}_p}(G)$ as:

$$\lambda_{\mathbb{E}_p}(G) = \left|\int_\Gamma \overline{z} f^*(z) dz\right| = \left|2i \int\int_G \frac{\partial}{\partial \overline{z}} (\overline{z} f^*(z))\, d\sigma\right| = \left|2i \int\int_G f^*(z) d\sigma\right|.$$

Since the boundary of the domain is analytic and $\overline{z}$ is real analytic on Γ, then by S. Ya. Khavinson's results on the regularity of extremal functions (see Theorem 5.13 in [19]) we know that $f^*(z)$ is analytic across Γ. Hence we can express $f^*(z)$ as the Cauchy integral of its boundary values, $f^*(z) = \frac{1}{2\pi i}\int_\Gamma \frac{f^*(w)}{w-z}dw$. Substituting this in the last equality, using Fubini's theorem, and bringing absolute values inside the integral we obtain

$$\begin{aligned}\lambda_{\mathbb{E}_p}(G) &= \left|\int\int_G \left(\frac{1}{\pi}\int_\Gamma \frac{f^*(w)}{w - z} dw\right)\, d\sigma\right| \\ &= \left|\int_\Gamma f^*(w) \left(\frac{1}{\pi}\int\int_G \frac{1}{w - z} d\sigma\right)\, dw\right| \\ &\leq \int_\Gamma |f^*(w)| \left|\frac{1}{\pi}\int\int_G \frac{1}{w - z} d\sigma\right|\, dw.\end{aligned}$$

Hölder's inequality yields

$$\lambda_{\mathbb{E}_p}(G) \le \|f^*\|_{\mathbb{L}_q(ds)} \left\| \frac{1}{\pi} \iint_G \frac{1}{w-z} d\sigma \right\|_{\mathbb{L}_p(ds,\Gamma)}$$
$$= \left\| \frac{1}{\pi} \iint_G \frac{1}{w-z} d\sigma \right\|_{\mathbb{L}_p(ds,\Gamma)}.$$

Let $F_G(z) = \frac{1}{\pi} \iint_G \frac{1}{w-z} d\sigma$, $z \in \mathbb{C}$. Now, the Ahlfor-Beurling estimate [2] (see [10] for a simple proof) implies that for a fixed $z \in \mathbb{C}$ and among domains with the same area, the function $|F_G(z)|$ attains its maximum value when the domain is a disk of radius ρ passing through z, which we denote by D_ρ. So, $|F_G(z)| \le |F_{D_\rho}(z)| \le \sqrt{\frac{A(D_\rho)}{\pi}} = \sqrt{\frac{A(G)}{\pi}}$. Therefore,

$$\lambda_{\mathbb{E}_p}(G) = \left\| \frac{1}{\pi} \iint_G \frac{1}{w-z} d\sigma \right\|_{\mathbb{L}_p(ds,\Gamma)}$$
$$= \left(\int_\Gamma \left| \frac{1}{\pi} \iint_G \frac{1}{w-z} d\sigma \right|^p ds \right)^{\frac{1}{p}}$$
$$\le \left(\int_\Gamma \left(\frac{A(G)}{\pi} \right)^{\frac{p}{2}} ds \right)^{\frac{1}{p}}$$
$$= \sqrt{\frac{A(G)}{\pi}} P(G)^{\frac{1}{p}}.$$

The theorem is proved. □

3. Characterization of disks and annuli in terms of approximations to $\overline{z}$ in $\mathbb{E}_p$ norm

PROPOSITION 3.1. *Let $p \ge 1$ and let $G = \{z \in \mathbb{C} : |z| < r\}$. Then:*
(i) The best approximation in $\mathbb{E}_p(G)$ to a general monomial of the type $\omega(z) = z^n \overline{z}^m$ for $m > n$ is the zero function. (For $m \le n$, it is clear that $z^n \overline{z}^m = r^{2m} z^{n-m}$ is its own best approximation.)
(ii) $\Lambda_p(G) = \|z^n \overline{z}^m\|_{\mathbb{L}_p(ds,\Gamma)} = \sqrt[p]{2\pi r^{p(n+m)+1}}$.
(iii) The best approximation to $\overline{z}$ in $\mathbb{E}_p(G)$ is the zero function and the $p-$analytic content of a disk of radius r is $\lambda_{\mathbb{E}_p}(G) = \|\overline{z}\|_{\mathbb{L}_p(ds,\Gamma)} = \sqrt[p]{2\pi r^{p+1}}$.

The proof is trivial, we only sketch it for the reader's convenience.

SKETCH OF PROOF. Let $G = \{z \in \mathbb{C} : |z| < r\}$, $p > 1$ and let $f(z) = \frac{|z^n \overline{z}^m|^p}{z^n \overline{z}^m} \frac{ds}{dz}$ for $m > n$. The function $f(z)$ annihilates $\overline{\mathbb{E}_p(G)}$ since, for $k \ge 0$

$$\int_\Gamma \left(\frac{|z^n \overline{z}^m|^p}{z^n \overline{z}^m} \right) z^k ds = \int_0^{2\pi} r^{(p-1)(n+m)+k+1} e^{i(k+m-n)\theta} d\theta = 0.$$

Set $f^*(z) = \frac{f(z)}{\|f\|_{\mathbb{L}_q(ds,\Gamma)}}$, so that $\|f^*\|_{\mathbb{L}_q(ds,\Gamma)} = 1$. Then with $g^*(z) = 0$ the extremality condition is satisfied:

$$f^*(z)\, z^n\overline{z}^m dz = \frac{\frac{|z^n\overline{z}^m|^p}{z^n\overline{z}^m}\frac{ds}{dz}}{\left\|\frac{|z^n\overline{z}^m|^p}{z^n\overline{z}^m}\right\|_q} z^n\overline{z}^m dz$$

$$= \frac{|z^n\overline{z}^m|^p}{\left\|(z^n\overline{z}^m)^{\frac{p}{q}}\right\|_q} ds.$$

For $p = 1$, let $f(z) = -iz^{m-n-1}\frac{ds}{dz}$ and $g^*(z) = 0$. Then $\int_\Gamma - iz^{m-n-1+k}dz = 0$, and $-iz^{m-n-1}(z^n\overline{z}^m)dz = r^{2m}d\theta$. Hence $f^*(z) = \frac{f(z)}{\|f(z)\|_{\mathbb{L}_1(ds,\Gamma)}}$ and $g^*(z)$ are both extremal.

Now, $\Lambda_p^p(G) = \|z^n\overline{z}^m\|^p_{\mathbb{L}_p(ds,\Gamma)} = \int_0^{2\pi}\left|r^{n+m}e^{i(n-m)\theta}\right|^p rd\theta = 2\pi r^{p(n+m)+1}$.

Taking $n = 0$ and $m = 1$ we obtain (iii). □

THEOREM 3.2. *Let G be a multiply connected bounded domain with the boundary consisting of n simple closed analytic curves. The zero function is the best approximation to $\overline{z}$ in $\mathbb{E}_p(G)$ if and only if G is a disk.*

PROOF. Necessity is obvious. For the converse, suppose that 0 is the best approximation to $\overline{z}$ in $\mathbb{E}_p(G)$. Then the extremality condition (2), for $p \geq 1$, can be written as

$$f^*(z)\overline{z}dz = const|z|^p ds$$

on each boundary component of the domain G. Without loss of generality we will assume the constant is positive. Dividing by z we can rewrite the equation above as

$$\frac{f^*(z)}{z}dz = const|z|^{p-2}ds. \tag{4}$$

Notice that $0 \in G$, otherwise $\int_\Gamma \frac{f^*(z)}{z}dz = 0$, yet $\int_\Gamma |z|^{p-2}ds \neq 0$ since this is a positive measure. For the same reason $f^*(0) \neq 0$, hence $\frac{f^*(z)}{z}$ has a pole at the origin.

Because the boundary of the domain is analytic, for each boundary component we can find a Schwarz function $S(z) = \overline{z}$, that is, a unique analytic function which at every point along the boundary component takes on the value $\overline{z}$ [14], [20]. Now, $(ds)^2 = dzd\overline{z} = S'(z)dz^2$, so $\frac{ds}{dz} = \sqrt{S'(z)}$ on Γ and we obtain that

$$\frac{f^*(z)}{z^{\frac{p}{2}}} = const\ S(z)^{\frac{p}{2}-1}\sqrt{S'(z)}.$$

Squaring both sides yields

$$\frac{[f^*(z)]^2}{z^p} = const\ \frac{d}{dz}\left[S(z)^{p-1}\right]. \tag{5}$$

This last equation implies that for each contour $S(z)^{p-1}$ is analytic throughout the domain, except at the origin.

We will now consider a few cases.

CASE 1. $p = 1$

When $p = 1$, $|f^*| \leq 1$ in G and $|f^*| = 1$ on Γ. Therefore, $f^*(z)$ is either a unimodular constant or the cover mapping of G onto the unit disk.

Suppose $f^*(z)$ is not constant. From Corollary 2.2 we have that $f^*(z) = e^{i\delta}\frac{|\bar{z}|ds}{\bar{z}dz}$ and $|f^*(z)| = 1$ almost everywhere on the boundary of G. By S. Ya. Khavinson's regularity results (Theorem 5.13 in [19]) $|f^*(z)| = 1$ everywhere on the boundary and $f^*(z)$ extends analytically across each boundary component. Therefore, $f^*(z)$ maps G onto the unit disk $\mathbb{D}$ taking each value in the disk k times, and wrapping each boundary component of Γ around the unit circle at least once, and always following the same positive direction. If that were not the case, and we suppose that at some point $w \in \Gamma$, $f^*(w)$ changes direction, at that point $\frac{df^*}{ds}(w) = 0$, so $\frac{df^*}{dz}(w) = \frac{df^*}{ds}\frac{ds}{dz} = 0$.

Hence, for z near w, $f^*(z) = f^*(w) + \mathcal{O}((z-w)^2)$. So $f^*(z)$ maps the "half" neighborhood of w that is in G onto a full neighborhood of $f^*(w)$, which means that $|f^*|$ can be greater than 1 near w, and that is a contradiction. Now, in order to wrap each boundary component of G around the circle, $f^*(z)$ has to go around the unit circle k times with $n \leq k$.

If we let $\Delta_{\arg} f^*(z)$ denote the change in the argument of $f^*(z)$ as z goes around the boundary of G, then $\Delta_{\arg} f^*(z) \geq n$. Moreover, the tangent vector to Γ traverses the boundary of G once in the clockwise direction, and $n-1$ times in the counterclockwise direction. Hence, remembering that Γ is analytic and $0 \in G$, by the argument principle we obtain that

$$\Delta_{\arg}\left(\frac{f^*(z)}{z}dz\right) = \Delta_{\arg} f^*(z) + \Delta_{\arg}\frac{1}{z} + \Delta_{\arg} dz \geq n - 1 + 2 - n = 1,$$

while $\Delta_{\arg}|z|^{p-2}ds = 0$ and from (4) we obtain a contradiction. Hence, for $p = 1$, $f^*(z)$ is a unimodular constant so from the equation preceding (4) we invoke that on Γ, $\bar{z}\frac{dz}{ds} = e^{ia}|z|$, where a is a real constant. Writing on each boundary component $z(s) = r(s)e^{ib(s)}$, substituting and separating real and imaginary parts yields $r' = \cos a$. Since each component is a closed curve, it cannot be a spiral, $\cos a$ must be zero, thus each component is a circle centered at the origin. Moreover, the case of the annulus is ruled out because $\frac{dz}{ds}$ changes directions between the two boundary circles, hence $|z| = const$ on Γ, and G is a disk centered at the origin.

CASE 2. $p > 1$, $p \notin \mathbb{N}$

If p is not an integer $S(z)^{p-1}$ may be multivalued. Yet, since the left hand side of (5) is $\mathcal{O}\left(\frac{1}{z^p}\right)$ near zero, it follows that $S(z)$ is $\mathcal{O}\left(\frac{1}{z}\right)$ in a neighborhood of the origin.

Also notice that if p is not an integer $S(z)$ cannot vanish anywhere in G. If it did it would be possible to obtain an unbounded singularity on the right hand side of (4) by differentiation, while the left hand side would remain bounded. Therefore the Schwarz function for every boundary component of G is analytic in the whole domain and has a simple pole at the origin. Moreover, since $\frac{[f^*(z)]^2}{z^p}$ remains the same when it is continued analytically throughout G, $S(z)$ has to be the same analytic function for each boundary component. So $S(z)^{p-1} = \int_\Gamma \frac{[f^*(z)]^2}{z^p}dz$ and from this we obtain that $S(z) = \frac{const}{z} + g(z)$, where $g(z)$ is analytic in G and is independent of which boundary component we consider. $S(z) = \bar{z}$ on the boundary. $S(z)z = |z|^2$ is real, positive on the boundary and analytic inside the domain G, hence it is constant. The boundary of the domain is therefore a circle centered at the origin.

CASE 3. $p > 1$, $p \in \mathbb{N}$

When p is even, i.e. $p = 2k$, (4) becomes

$$\frac{f^*(z)}{z^k} dz = const|z|^{k-1} ds, \tag{6}$$

which in turn yields

$$\frac{[f^*(z)]^2}{z^{2k}} = const \frac{d}{dz} \left[S(z)^{2k-1}\right]. \tag{7}$$

(6) implies that $S(z)^{2k-1}$ is analytic throughout G and has a pole of order $2k-1$ at the origin in G, so $S(z)$ has to have a simple pole at the origin. Following the same reasoning as in case 2 we can conclude that $S(z)^{2k-1}$ is the same for every boundary component.

Then $\overline{z}^{2k-1} = S(z)^{2k-1} = \frac{const}{z^{2k-1}} + g(z)$ for $g(z)$ analytic in G. Multiplying through by z^{2k-1} we have once again that $|z| = const$.

For p odd, i.e. $p = 2k+1$, (5) can be written as

$$\frac{[f^*(z)]^2}{z^{2k+1}} = const \frac{d}{dz} \left[S(z)^{2k}\right] \tag{8}$$

So $S(z)^{2k} = \frac{const}{z^{2k}} + g(z)$, with $g(z)$ analytic in G. Hence, once more, the boundary of the domain is a circle. □

DEFINITION 3.3. ([6], Ch. 10) *Let G be a Jordan domain with rectifiable boundary Γ, let $z = \phi(w)$ map G onto $|w| < 1$. Since $\phi' \in H^1$ and has no zeros, it has a canonical factorization $\phi'(w) = S(w)\Phi(w)$ where S is a singular inner function and Φ is an outer function. G is said to satisfy the Smirnov condition if $S(w) = 1$, i.e. if ϕ' is purely outer.*

It is the case that G is a Smirnov domain if and only if $\mathbb{E}^p(G)$ coincides with the $\mathbb{L}^p(\Gamma)$ closure of the polynomials. We will use repeatedly the property that if a function $f \in \mathbb{E}^p(G)$ belongs to $\mathbb{L}^p(\Gamma)$ with $q > p$, then $f \in \mathbb{E}^q(G)$.

REMARK 3.4. For a simply connected domain we can significantly relax the assumption of analyticity of the boundary in Theorem 3.2 and obtain that the domain is a disk invoking the following result from [8].

THEOREM 3.5. *(Thm. 3.29 in [8]). Let G be a Jordan domain in $\mathbb{R}^2 \cong \mathbb{C}$ containing 0 and with the rectifiable boundary Γ satisfying the Smirnov condition. Suppose the harmonic measure on Γ with respect to 0 equals $c|z|^\alpha ds$ for $z \in \Gamma$, where ds denotes arclength measure on Γ, $\alpha \in R$ and c is a positive constant. Then*
(i) For $\alpha = -2$, the solutions are precisely all disks G containing 0.
(ii) For $\alpha = -3, -4, -5, \ldots$ there are solutions G which are not disks.
(iii) For all other values of α, the only solutions are disks centered at 0.

To apply this result in our context we need first to notice that the positive measure $\frac{f(z)}{z}dz = const|z|^{p-2}\,ds$ annihilates all analytic functions vanishing at the origin and hence is, after normalizing by a scalar multiple, a representing measure for analytic functions at the origin. Moreover, since the domain is simply connected, we can separate real and imaginary parts and then conclude that this latter measure is precisely the harmonic measure at 0. Because $p-2 \geq -1$, part (iii) applies and the domain is a disk centered at the origin.

THEOREM 3.6. *Let $p, q \geq 1$, $\frac{1}{p}+\frac{1}{q}=1$.*
Let G be an annulus $\{z : 0 < r < |z| < R\}$ and $\Gamma = \gamma_1 \cup \gamma_2$ be its boundary.
(i) For $p>1$ the best analytic approximation to $\omega = z^n\overline{z}^m$ in $\mathbb{E}_p(G)$ is unique and equal to $g^*(z) = cz^{n-m}$ where $c(n,m,p) = \frac{r^{2m+q(n-m)+\frac{q}{p}}+R^{2m+q(n-m)+\frac{q}{p}}}{r^{q(n-m)+\frac{q}{p}}+R^{q(n-m)+\frac{q}{p}}}$.
(ii) For $p=1$, and $n-m=-1$, the set of functions that are closest to $\omega = z^n\overline{z}^m$ in $\mathbb{E}_1(G)$ consist of all functions of the form $g^*(z) = cz^{n-m}$ where c is any constant such that $r^{2m} \leq c \leq R^{2m}$.
(iii) For $p>1$, the distance from $z^n\overline{z}^m$ to $\mathbb{E}_p(G)$ is

$$\begin{aligned}\Lambda_p(G) &:= \|z^n\overline{z}^m - g^*(z)\|_{\mathbb{L}_p(ds,\Gamma)}\\ &= \frac{(rR)^{n-m}\left(R^{2m}-r^{2m}\right)}{r^{q(n-m)+\frac{q}{p}}+R^{q(n-m)+\frac{q}{p}}}\sqrt[p]{2\pi\left[\left(r^{\frac{1}{p}}R^{\frac{q}{p}(1-n-m)}\right)^p + \left(R^{\frac{1}{p}}r^{\frac{q}{p}(1-n-m)}\right)^p\right]}.\end{aligned}$$

For $p=1$, $\Lambda_1(G) = \left|\int_\Gamma z^n\overline{z}^m f^*(z)dz\right| = 2\pi(R^{2m}+r^{2m})$.

Note: For $n-m \neq -1$, we have been unable to find the best approximation in closed form, see the remark at the end of the proof.

PROOF. Consider $f(z) = \frac{|z^n\overline{z}^m - cz^{n-m}|^p}{z^n\overline{z}^m - cz^{n-m}}\frac{ds}{dz}$. Then

$$\begin{aligned}&\int_\Gamma \left(\frac{|z^n\overline{z}^m - cz^{n-m}|^p}{z^n\overline{z}^m - cz^{n-m}}\right) z^k ds\\ &= \int_0^{2\pi} \frac{\left|R^n e^{in\theta} R^m e^{-im\theta} - cR^{n-m}e^{i(n-m)\theta}\right|^p}{R^n e^{in\theta} R^m e^{-im\theta} - cR^{n-m}e^{i(n-m)\theta}} R^{k+1}e^{ik\theta}\,d\theta\\ &+ \int_0^{2\pi} \frac{\left|r^n e^{in\theta} r^m e^{-im\theta} - cr^{n-m}e^{i(n-m)\theta}\right|^p}{r^n e^{in\theta} r^m e^{-im\theta} - cr^{n-m}e^{i(n-m)\theta}} r^{k+1}e^{ik\theta}\,d\theta\\ &= \left(R^{\frac{p(n-m)+q(k+1)}{q}}\frac{|R^{2m}-c|^p}{R^{2m}-c} + r^{\frac{p(n-m)+q(k+1)}{q}}\frac{|r^{2m}-c|^p}{r^{2m}-c}\right)\int_0^{2\pi} e^{i(k+m-n)\theta}\,d\theta\\ &= 0, \text{ unless } k = n-m.\end{aligned}$$

and

$$2\pi\left(R^{p(n-m)+1}\frac{|R^{2m}-c|^p}{R^{2m}-c} + r^{p(n-m)+1}\frac{|r^{2m}-c|^p}{r^{2m}-c}\right) = 0$$

if

$$R^{p(n-m)+1}\frac{|R^{2m}-c|^p}{R^{2m}-c} = -r^{p(n-m)+1}\frac{|r^{2m}-c|^p}{r^{2m}-c},$$

which is only possible if $r^{2m} < c < R^{2m}$. In that case,

$$\left(\frac{c - r^{2m}}{R^{2m} - c}\right)^{p-1} = \left(\frac{R}{r}\right)^{p(n-m)+1}$$

and after some algebra we obtain that

$$c = \frac{r^{2m+q(n-m)+\frac{q}{p}} + R^{2m+q(n-m)+\frac{q}{p}}}{r^{q(n-m)+\frac{q}{p}} + R^{q(n-m)+\frac{q}{p}}}.$$

Therefore, $f(z)$ annihilates $\overline{\mathbb{E}_p(G)}$.
Now let $f^*(z) = \frac{f}{\|f\|_{\mathrm{L}_q(ds)}}$ so that $\|f^*\|_{\mathrm{L}_q(ds)} = 1$ and let $g^*(z) = cz^{n-m}$. Then,

$$\begin{aligned} f^*(z)(z^n\overline{z}^m - g^*(z))dz &= \frac{\frac{|z^n\overline{z}^m - cz^{n-m}|^p}{z^n\overline{z}^m - cz^{n-m}}}{\left\|\frac{|z^n\overline{z}^m - cz^{n-m}|^p}{z^n\overline{z}^m - cz^{n-m}}\right\|_q}(z^n\overline{z}^m - cz^{n-m})ds \\ &= \frac{|z^n\overline{z}^m - cz^{n-m}|^p}{\left\|(z^n\overline{z}^m - cz^{n-m})^{\frac{p}{q}}\right\|_q}ds \\ &= \frac{|z^n\overline{z}^m - cz^{n-m}|^p}{\|z^n\overline{z}^m - cz^{n-m}\|_p^{p-1}}ds, \end{aligned}$$

which is condition *(iii)* in Corollary 2.2. Therefore $f^*(z)$ and $g^*(z)$ are extremal.
For $p = 1$, and $n - m = -1$, by Corollary 2.2, $f^*(z)$ and $g^*(z)$ are extremal if and only if they satisfy that $f^*(z)(z^n\overline{z}^m - g^*(z))dz = |z^n\overline{z}^m - g^*(z)|ds$ on each boundary component of the annulus.
Consider $f^*(z) = -i$ and $g^*(z) = \frac{c}{z}$.
On $\gamma_1 = \{z \in \mathbb{C} : |z| = r\}$, with clockwise orientation on the boundary we have

$$-i(z^n\overline{z}^m - \frac{c}{z})dz = -(r^{2m} - c)d\theta$$

and on the other hand

$$|z^n\overline{z}^m - \frac{c}{z}|ds = \left|r^{2m} - c\right| d\theta$$

The same analysis on $\gamma_2 = \{z \in \mathbb{C} : |z| = R\}$, where the orientation on the boundary is counterclockwise, yields

$$-i(z^n\overline{z}^m - \frac{c}{z})dz = (R^{2m} - c)d\theta$$

and

$$|z^n\overline{z}^m - \frac{c}{z}|ds = \left|R^{2m} - c\right| d\theta.$$

Which means

$$-(r^{2m} - c)d\theta = \left|r^{2m} - c\right| d\theta$$

and

$$(R^{2m} - c)d\theta = \left|R^{2m} - c\right| d\theta.$$

These two equations hold simultaneously for any constant c in the interval $[r^{2m}, R^{2m}]$.
Finally we compute $\Lambda_p(G)$.

For $p>1$, recalling that $c(n,m,p)=\frac{r^{2m+q(n-m)+\frac{q}{p}}+R^{2m+q(n-m)+\frac{q}{p}}}{r^{q(n-m)+\frac{q}{p}}+R^{q(n-m)+\frac{q}{p}}}$, we have

$$\begin{aligned}
\Lambda_p^p(G) &= \left\|z^n\overline{z}^m - cz^{n-m}\right\|^p_{\mathbb{L}_p(ds,\Gamma)} \\
&= 2\pi\left[r^{p(n-m)+1}\left(c-r^{2m}\right)^p + R^{p(n-m)+1}\left(R^{2m}-c\right)^p\right] \\
&= 2\pi\left[\left(r^{\frac{1}{p}}R^{\frac{q}{p}(1-n-m)}\right)^p + \left(R^{\frac{1}{p}}r^{\frac{q}{p}(1-n-m)}\right)^p\right]\left(\frac{(rR)^{n-m}\left(R^{2m}-r^{2m}\right)}{r^{q(n-m)+\frac{q}{p}}+R^{q(n-m)+\frac{q}{p}}}\right)^p.
\end{aligned}$$

Therefore $\Lambda_p(G)=\frac{(rR)^{n-m}(R^{2m}-r^{2m})}{r^{q(n-m)+\frac{q}{p}}+R^{q(n-m)+\frac{q}{p}}}\sqrt[p]{2\pi\left[\left(r^{\frac{1}{p}}R^{\frac{q}{p}(1-n-m)}\right)^p+\left(R^{\frac{1}{p}}r^{\frac{q}{p}(1-n-m)}\right)^p\right]}$.

Now, for $p=1$ and $n-m=-1$

$$\begin{aligned}
\Lambda_1(G) &= \left|\int_\Gamma z^n\overline{z}^m f^*(z)dz\right| \\
&= \left|\int_0^{2\pi} r^{2m}+R^{2m}d\theta\right| \\
&= 2\pi(r^{2m}+R^{2m}).
\end{aligned}$$

The proof of Theorem 3.6 is now complete. □

REMARK 3.7. When $p=1$ and $n-m\neq -1$, because the boundary is analytic and $f^*(z)$ is continuous on $\overline{G}$, $|f^*(z)|=1$ everywhere on the boundary. Therefore, $f^*(z)$ is either constant or a $k-$sheeted covering of the unit disk. It is not a constant since $\int_\Gamma z^n\overline{z}^m dz=0$ unless $n-m=-1$. So $f^*(z)$ maps G onto a $k-$sheeted cover of the unit disk with $k\geq n$. Hence the best approximation to $z^n\overline{z}^m$ cannot be a monomial cz^{n-m}. Moreover, it follows from the duality relations that $f^*(z)$ has to be a transcendental function.

By letting $n=0$ and $m=1$ we have the following corollary.

COROLLARY 3.8. *Let* $\frac{1}{p}+\frac{1}{q}=1$.
Let G *be an annulus* $\{z : 0<r<|z|<R\}$.
(i) For $p>1$ *the best analytic approximation to* $\omega=\overline{z}$ *in* $\mathbb{E}_p(G)$ *is* $g^*(z)=\frac{rR}{z}$.
(ii) For $p=1$, *all functions* $g^*(z)=\frac{c}{z}$ *for any constant* $c\in[r^2,R^2]$, *serve as the best approximation to* $\overline{z}$ *in* $\mathbb{E}_1(G)$.
(iii) For $p\geq 1$ *the* $p-$*analytic content of* G *is* $\lambda_{\mathbb{E}_p}(G)=(R-r)(2\pi(R+r))^{\frac{1}{p}}$

Notice that the best approximation to $\overline{z}$ in $\mathbb{E}_p(G)$ is $g^*(z)=\frac{rR}{z}$ independent of p!

Next we will prove a partial converse for Theorem 3.6 in the case when $p=1$. For that we will need the following lemma.

LEMMA 3.9. *Let G be a multiply connected domain in $\mathbb{C}$ with analytic boundary consisting of n components. If $g^*(z) = \frac{c}{z}$ is the best approximation to $\overline{z}$ in $\mathbb{E}_1(G)$ and $\overline{z}$ does not coincide with $\frac{c}{z}$ on any of the boundary components then:*
i) $f^*(z)$, the extremal function in $\mathbb{E}^1_\infty(G)$ for which $\sup_{f\in\mathbb{E}^1_\infty(G)} \left|\int_\Gamma \overline{z} f(z)dz\right|$ is attained, is a unimodular constant.
ii) The number n of boundary components of G is 2.

PROOF. Replicating the argument used in Theorem 3.2, case 1, we can show that unless $f^*(z)$ is a constant, it is a $k-$sheeted covering of the unit disk, with $k \geq n$, thus $\Delta_{\arg} f^*(z) \geq n$. Moreover, the tangent vector to Γ goes along the boundary of G once in the clockwise direction, and $n-1$ times in the counterclockwise direction. So $\Delta_{\arg} f^*(z)dz \geq n+2-n = 2$. Now, since the boundary of the domain is analytic and we are assuming that $\frac{1}{z}$ is analytic in G, $\frac{1}{z}$ has no poles in G. By the argument principle we can say that

$$\Delta_{\arg}\left(\overline{z} - \frac{c}{z}\right) = \Delta_{\arg}\frac{|z|^2 - c}{z} = \Delta_{\arg}(|z|^2 - c) + \Delta_{\arg}\frac{1}{z} = 0.$$

So $\Delta_{\arg}(\overline{z} - \frac{c}{z})f^*dz = \Delta_{\arg} f^*(z)dz + \Delta_{\arg}\left(\overline{z} - \frac{c}{z}\right) \geq 2$. Yet Corollary 2.2 (*iii*) yields that $\Delta_{\arg}(\overline{z} - \frac{c}{z})f^*dz = 0$ since it has constant argument on Γ, so we have a contradiction. Hence, $f^*(z)$ has to be constant.

With $f^*(z)$ constant we have that $\Delta_{\arg}(\overline{z} - \frac{c}{z})f^*dz = \Delta_{\arg}dz = 2 - n = 0$ therefore the number of boundary components of G is $n = 2$. □

THEOREM 3.10. *Let G be a multiply connected domain in $\mathbb{C}$ with analytic boundary Γ. If $g^*(z) = \frac{c}{z}$ is the best approximation to $\overline{z}$ in $\mathbb{E}_1(G)$ and the hypotheses of Lemma 3.9 are satisfied, then G is an annulus.*

PROOF. (That the best analytic approximation to $\overline{z}$ in $\mathbb{E}_1$ of the annulus is $g^*(z) = \frac{c}{z}$ follows from Corollary 3.8.)
We infer from Lemma 3.9 that $f^*(z) = e^{i\alpha}$. By the duality relations we obtain

$$\lambda_{\mathbb{E}_1} = \left|\int_\Gamma \overline{z}dz\right| = \left|\int_\Gamma (\overline{z} - \frac{c}{z})dz\right| \leq \int_\Gamma \left|\overline{z} - \frac{c}{z}\right| ds = \lambda_{\mathbb{E}_1}. \quad (9)$$

Therefore equality holds throughout.
Now, since $|z|^2 - c$ is real and the boundary is analytic, (9) implies that $\arg\left(\frac{dz}{z}\right)$ is constant on every boundary component of G. On the other hand we have from Lemma 3.9 (*ii*) that G has two boundary components γ_1 and γ_2, with opposite orientation. So letting $z(s) = r(s)e^{ib(s)}$, with s being the arclength parameter, since $\left|\frac{dz}{ds}\right| = 1$, by differentiating we obtain

$$\frac{dz}{ds} = (ir(s)b'(s) + r'(s))e^{ib(s)} = e^{ia_j + ib(s)}, \ j = 1,2$$

were a_j, $j = 1,2$ are constants on γ_1 and γ_2 respectively. This yields that

$$ir(s)b'(s) + r'(s) = e^{ia_j}, \ j = 1,2.$$

Differentiating again, we obtain

$$r''(s) + i(r(s)b'(s))' = 0,$$

hence ($r(s)$ and $b(s)$ are real-valued functions) $r''(s) = 0$ and $r(s)$ is a linear function. Recalling that the boundary of the domain consists of two closed curves we conclude that $r(s)$ is linear and periodic, hence it is constant on each boundary component. So the boundary of the domain consists of two concentric circles and the domain is an annulus. □

REMARK 3.11. In Lemma 3.9, if $\overline{z}$ does coincide with $\frac{c}{z}$ on one of the boundary components, say γ_o, i.e. if that component is a circle, then on that boundary component $|f^*| \leq 1$ while on the remaining components $|f^*| = 1$. In this case we can only infer that $\underset{\Gamma\setminus\gamma_o}{\Delta \arg} f^* \geq n - 1$ and the argument above fails. We conjecture that Theorem 3.10 holds for all $p \geq 1$ and without the additional hypothesis in Lemma 3.9. Yet, we have not been able to prove it.

4. The Bergman Space case: Characterization of disks and annuli in terms of the best analytic approximation to $\overline{z}$ in $\mathbb{A}_p$ norm.

Let $d\sigma$ be area measure on G.

We use the standard notation $\mathbb{W}^{1,q}(G)$ and $\mathbb{W}^{1,q}_o(G)$ for Sobolev spaces and Sobolev spaces with vanishing boundary values. The reader may consult [9, Ch.5], [1] for details.

Khavin's lemma (see [20]) describes the annihilator of $A_p(G)$ as follows:
For $p > 1$,

$$\begin{aligned} Ann(\mathbb{A}_p(G)) &:= \left\{ f \in \mathbb{L}_q(d\sigma, G) : \int_G fg d\sigma = 0 \text{ for all } g \in \mathbb{A}_p(G) \right\} \\ &= \left\{ \frac{\partial u}{\partial \overline{z}},\ u \in \mathbb{W}^{1,q}_o(G) \right\}. \end{aligned}$$

For $p = 1$,

$$Ann(\mathbb{A}_1(G)) := \left\{ weak(*) \text{ closure of } \frac{\partial u}{\partial \overline{z}},\ u \in \mathbb{W}^{1,\infty}(G), \text{ in } \mathbb{L}_\infty(d\sigma, G) \right\}.$$

DEFINITION 4.1. *The Bergman $p-$analytic content of a domain G is*

$$\lambda_{\mathbb{A}_p}(G) := \inf_{g \in \mathbb{A}_p(G)} \|\overline{z} - g(z)\|_{\mathbb{L}_p(d\sigma, G)}.$$

By the Hahn-Banach theorem,

$$\lambda_{\mathbb{A}_p}(G) = \max_{f \in Ann(\mathbb{A}_p(G)), \|f\| \leq 1} \left| \int_G \overline{z} f d\sigma \right|.$$

A similar result to Corollary 2.2 holds in the context of Bergman spaces; we state it as Corollary 4.2 for completeness. See [17] Theorem 3.1, Remarks (i) and (iv). Also see [18] p. 940.

COROLLARY 4.2. *Let $\frac{1}{p} + \frac{1}{q} = 1$, and let $\omega(z) \in \mathbb{L}_p(d\sigma, G)$. Then the following hold:*

(i) $\inf_{g \in \mathbb{A}_p(G)} \|\omega(z) - g(z)\|_{\mathbb{L}_p(d\sigma,G)} = \sup_{f \in Ann(\mathbb{A}_p(G)), \|f\| \leq 1} \left| \int_G \omega(z) f d\sigma \right|.$

(ii) There exist extremal functions $g^*(z) \in \mathbb{A}_p(G)$ and $f^*(z) \in Ann(\mathbb{A}_p(G))$ for which the infimum and the supremum are attained in (i).

(iii) When $p > 1$, $g^*(z) \in \mathbb{A}_p(G)$ and $f^*(z) \in Ann(\mathbb{A}_p(G))$ are extremal if and only if, for some real number δ,

$$\begin{aligned} e^{i\delta} f^*(z)(\omega(z) - g^*(z)) &\geq 0 \text{ in } G, \\ \Lambda^p_{\mathbb{A}_p} |f^*(z)|^q &= |\omega(z) - g^*(z)|^p \text{ in } G, \end{aligned}$$

where $\Lambda_{\mathbb{A}_p} = \|\omega(z) - g^*(z)\|_{\mathbb{L}_p(d\sigma,G)}$. When $p = 1$ the conditions above become $e^{i\delta} f^*(z)(\omega(z) - g^*(z)) = |\omega(z) - g^*(z)|$ a.e. in G.

(iv) For $p > 1$ the best approximations $g^*(z) \in \mathbb{A}_p(G)$ and $f^*(z) \in Ann(\mathbb{A}_p(G))$ are always unique. For $p = 1$ and $\omega(z)$ continuous in G, the best approximation $g^*(z) \in \mathbb{A}_p(G)$ is unique. For discontinuous $\omega(z)$ the best approximation need not be unique. Also, in the case where $p = 1$ the duality condition in (iii) implies that $f^*(z) \in Ann(\mathbb{A}_1(G))$ is unique, up to a unimodular constant, provided that $\omega(z)$ does not coincide with an analytic function on a set of positive area measure.

REMARK 4.3. For the case of the disk $\mathbb{D} = \{z \in \mathbb{C} : |z| < r\}$ it was shown in [17] Proposition 2.3, that the best rational approximation in $\mathbb{A}_p(\mathbb{D})$ to $\omega = z^n \bar{z}^m$ for $p \geq 1$ and $m > n$ is $g^*(z) = 0$. When $m \leq n$, $g^*(z) = cz^{n-m}$, where $c = c(n, m, p)$ is an appropriate constant.

In that case we can compute the Bergman $p-$analytic content of $\mathbb{D}$ as follows:

$$\begin{aligned} \lambda_{\mathbb{A}_p}(\mathbb{D}) &= \sqrt[p]{\int_0^{2\pi} \int_0^r |te^{-it}|^p \, t dt d\theta} \\ &= \sqrt[p]{2\pi \int_0^r |t|^p \, t dt} \\ &= \sqrt[p]{\frac{2\pi r^{p+2}}{p+2}}. \end{aligned}$$

Following the argument in [17] we find the extremal functions for the case of the annulus.

PROPOSITION 4.4. *Let $p, q \geq 1$, $\frac{1}{p} + \frac{1}{q} = 1$. Let G be an annulus $\{z : r < |z| < R, \ r < R\}$. The best analytic approximation to $\omega = z^n \bar{z}^m$ in $\mathbb{A}_p(G)$ is $g^*(z) = cz^{n-m}$, with $c = c(n, m, p)$ satisfying*

$$\int_r^R t^{p(n-m)+1} \left| t^{2m} - c \right|^{p-1} sgn(t^{2m} - c) dt = 0$$

and $\Lambda_{\mathbb{A}_p}(G) = \sqrt[p]{2\pi \int_r^R \left| t^{2m} - c \right|^p t^{p(n-m)+1} dt}.$

In particular, if $n = 0$ *and* $m = 1$*, for* $p \geq 1$*, the Bergman* $p-$*analytic content of* G *in* $\mathbb{A}_p$ *is* $\lambda_{\mathbb{A}_p}(G) = \sqrt[p]{2\pi \int_r^R \left|c - t^2\right|^p t^{-\frac{p}{q}} dt}$ *where* $c(0,1,p)$ *is such that* $\int_r^R t^{1-p} \left|t^2 - c\right|^{p-1} sgn(t^2 - c) dt = 0.$

PROOF. Consider $f(z) = \frac{|z^n\overline{z}^m - cz^{n-m}|^p}{z^n\overline{z}^m - cz^{n-m}}$,

$$\begin{aligned}\int_G \left(\frac{|z^n\overline{z}^m - cz^{n-m}|^p}{z^n\overline{z}^m - cz^{n-m}}\right) z^k d\sigma &= \int_0^{2\pi}\int_r^R \frac{|z^n\overline{z}^m - cz^{n-m}|^p}{z^n\overline{z}^m - cz^{n-m}} z^k t dt d\theta \\ &= \int_r^R \frac{t^{\frac{p}{q}(n-m)+k+1}\left|t^{2m} - c\right|^p}{t^{2m} - c}\left(\int_0^{2\pi} e^{-i(k+(n-m))\theta} d\theta\right) dt \\ &= 0, \text{ unless } k = n - m.\end{aligned}$$

If we choose $c = c(p)$, so that $\int_r^R t^{p(n-m)+1}\left|t^{2m} - c\right|^{p-1} sgn(t^{2m} - c) dt = 0$, then $f(z) \in Ann(\mathbb{A}_p(G))$.

Defining $f^*(z) = \frac{f}{\|f\|_{\mathbb{L}_q(d\sigma,G)}}$ we can check that necessary and sufficient (Corollary 4.2 (*iii*)) conditions for extremality hold and the result follows.

To compute $\Lambda^p_{\mathbb{A}_p}(G)$:

$$\Lambda^p_{\mathbb{A}_p}(G) := \int_G \left|z^n\overline{z}^m - cz^{n-m}\right|^p d\sigma = \int_0^{2\pi}\int_r^R \left|t^{n+m}e^{i(n-m)\theta} - ct^{n-m}e^{i(n-m)\theta}\right|^p t dt d\theta$$

$$= \int_0^{2\pi}\int_r^R t^{p(n-m)+1}\left|t^{2m} - c\right|^p dt d\theta = 2\pi \int_r^R t^{p(n-m)+1}\left|t^{2m} - c\right|^p dt.$$

□

THEOREM 4.5. *Let* G *be a bounded domain with analytic boundary. The best analytic approximation to* $\overline{z}$ *in* $\mathbb{A}^p(G)$ *is* $g^*(z) = 0$ *if and only if* G *is a disk.*

PROOF. If G is a disk, the best rational approximation to $\overline{z}$ *in* $\mathbb{A}_p(G)$ is $g^*(z) = 0$ by Remark 4.3.

Now suppose 0 is the best approximation to $\overline{z}$ in $\mathbb{A}^p(G)$. First assume $p > 1$. In this case Corollary 4.2 (*iii*) can be written as $|z|^p = \lambda^p |f^*|^q$, so $f(z) = \frac{|\overline{z}|^p}{\overline{z}}$ annihilates $\overline{\mathbb{A}^p(G)}$, $f^*(z) = \frac{f(z)}{\|f(z)\|_q}$, and by Khavin's Lemma $f^*(z) = \frac{\partial u}{\partial \overline{z}}$ for some $u \in \mathbb{W}^{1,q}_o(G)$. Hence,

$$\frac{\partial u}{\partial \overline{z}} = const \frac{|\overline{z}|^p}{\overline{z}}.$$

Integrating with respect to $\overline{z}$ we obtain

$$u(z) = \int \frac{\partial u}{\partial \overline{z}} d\overline{z} = const \int z^{\frac{p}{2}} \overline{z}^{\frac{p}{2}-1} d\overline{z} = const|z|^p + h(z),$$

where $h(z)$ is analytic.

Since $u(z) \in \mathbb{W}^{1,q}_o(G)$ and $|z|^p$ is real analytic near Γ, it is easy to see that for any sequence of domains G_j, $\cup G_j = G$ with rectifiable boundaries Γ_j, $\|u\|_{\mathbb{L}_q(\Gamma_j, ds)}$ are bounded, so $h(z) \in \mathbb{E}_q(G)$. Now, $u(z) = 0$ a.e. on Γ, hence $h(z) = -|z|^p$ a.e. on Γ. Since Γ is analytic, and $h(z) \in \mathbb{E}_q(G)$ for $q \geq 1$ and has bounded boundary values, $h(z)$ is bounded in G. But $h(z)$ has real boundary values on Γ a.e., hence $h(z)$ is constant. Thus $|z|^p$ is constant a.e. on Γ, so G is a disk centered at the origin. The case $p = 1$ and $q = \infty$ requires only small modifications that are left to the reader. □

Along the same lines we also have:

THEOREM 4.6. *Let G be a finitely connected domain with analytic boundary, $p \geq 1$. G is an annulus centered at the origin if and only if the best analytic approximation to $\bar{z}$ in $\mathbb{A}^p(G)$ is $g^*(z) = \frac{c}{z}$.*

PROOF. If G is an annulus we have already seen in Proposition 4.4 that the best rational approximation to $\bar{z}$ in $\mathbb{A}_p(G)$, $p \geq 1$, is $g^*(z) = \frac{c}{z}$.

To prove the converse, suppose the best approximation to $\bar{z}$ in $\mathbb{A}^p(G)$ is $g^*(z) = \frac{c}{z}$. Once again for the sake of clarity we focus on the case $p > 1$, the remaining case only requires small modifications that are left to the reader. Corollary 4.2 (iii) yields that $f^*(z) = \frac{|\bar{z}-\frac{c}{z}|^p}{\bar{z}-\frac{c}{z}} \in Ann(\overline{\mathbb{A}^p(G)})$, and Khavin's Lemma yields that $f^*(z) = \frac{\partial u}{\partial \bar{z}}$ for some $u \in \mathbb{W}_o^{1,q}(G)$.
Denoting $|z|$ by r we have

$$\frac{\partial u}{\partial \bar{z}} = \frac{|\bar{z}-\frac{c}{z}|^p}{\bar{z}-\frac{c}{z}} = \frac{|\bar{z}z-c|^p}{\bar{z}z-c}\frac{z}{z^{\frac{p}{2}}\bar{z}^{\frac{p}{2}}} = \frac{|r^2-c|^{p-1}}{r^{p-2}\bar{z}}sign(r^2-c).$$

Integrating we have that

$$\begin{aligned}(10)\qquad \int \frac{\partial u}{\partial \bar{z}}d\bar{z} &= \int \frac{|r^2-c|^{p-1}}{r^{p-2}\bar{z}}sign(r^2-c)d\bar{z}\\ &= \int \frac{(r^2-c)^{p-1}}{r^{p-2}}sign(r^2-c)\frac{d}{d\bar{z}}\log|z\bar{z}|d\bar{z}\\ &= 2\int \frac{(r^2-c)^{p-1}}{r^{p-2}}sign(r^2-c)d\log r.\end{aligned}$$

Since $0 \notin G$, this integral is bounded away from zero and yields a real-valued function $F(r)$ for all $r > 0$. So $u(z) = F(r) + h(z)$ with $h(z)$ analytic.

As in the proof of Theorem 4.5, $h(z)$ extends across the boundary and hence belongs to $\mathbb{H}^\infty(G)$. Now, because $u(z) \in \mathbb{W}_o^{1,q}(G)$, $u(z) = F(r) + h(z) = 0$ a.e. on the boundary of G. So $h(z)$ is real valued almost everywhere on the boundary. Hence it has to be real inside the domain as well and therefore constant.

Now note that $u = F(r) + const$ and $u = 0$ on Γ. Moreover from (10) it readily follows that $F'(r) = 0$ only at one point $r_o = \sqrt{c}$, where F' changes sign. Hence F may take the same value only twice, so Γ consists of two components and on each one the value of $F(r)$ is the same, i.e. Γ consists of two concentric arcs centered at the origin. Since G is bounded and $0 \notin G$ ($\frac{1}{z}$ is analytic in G!), G must be an annulus. □

5. Final Remarks

For the Bergman norm, assuming that G is a multiply connected domain with analytic boundary, we were able to prove that for all $p \geq 1$ the domain is an annulus whenever the best approximation to $\overline{z}$ is $\frac{c}{z}$ (and that the domain is a disk, whenever the best approximation to $\overline{z}$ is a constant function). Our proof relies on the assumption of analyticity of the boundary. However, it is easy to see from the proof that this assumption can be relaxed and we only need assume that the domain G is Smirnov. We do not know whether the result holds for domains with arbitrary rectifiable boundaries.

The Smirnov norms case turns out to be more difficult. One of the reasons is that knowing the best approximation in Bergman norm determines the extremal function in the dual problem throughout the domain (although in a vast set $Ann(\mathbb{A}_p)$). In the $\mathbb{E}_p$ setting, it only determines the extremal function in the dual problem, although analytic in the domain, on parts of the boundary where $\overline{z}$ does not coincide with its best approximation, which unfortunately could happen a priori. In the Bergman setting this can never happen because two real analytic functions can never coincide on a set of positive area measure without being identical.

If a constant is the best approximation to $\overline{z}$ in $\mathbb{E}_p$, we showed in Theorem 3.2, for multiply connected domains and under the assumption of analyticity of the boundary, that the domain is a disk. We were able to reach the same conclusion for Jordan domains with rectifiable boundaries satisfying the Smirnov condition, but only when the domain is simply connected. We think it should be possible to generalize Theorem 3.2 to multiply connected domains with weaker regularity conditions imposed on the boundary.

The following question seems natural in connection with Remark 3.4 (and Thm. 3.2). Let G be a finitely connected domain containing the origin and assume that

$(*)$ *the measure* $const|z|^\alpha ds$, $\alpha \in R$, *on the boundary* Γ *is a representing measure at the origin for analytic functions in* G, *say, continuous in* $\overline{G}$.

Does condition $(*)$ alone imply that G must be simply connected? If so, (cf. Remark 3.4) then for $\alpha > -2$, G must be a disk centered at the origin. Perhaps, condition $(*)$ implies that G is simply connected only for specific values of α, what are these values and what happens in the remaining cases? Under a less restrictive regularity assumption, say assuming the boundary of G merely rectifiable, even for $\alpha = 0$, there exist highly nonregular, non-Smirnov domains, so called pseudocircles, for which $(*)$ still holds (cf. [6], Ch. 10).

When the domain is an annulus we found the best $\mathbb{E}_p$-approximation to any monomial $z^n\overline{z}^m$ explicitly for all $p > 1$ and for $p = 1$ when $n - m = -1$. Yet, when $p = 1$ and $n - m \neq -1$, the extremal function f^* in the dual problem is a trascendental function, hence we can only conclude that the best approximation to $z^n\overline{z}^m$ is not a monomial. It would be worthwhile to study the best approximation of such monomials in $\mathbb{E}_1$ of the annulus in greater detail.

In Theorem 3.10 we show for $p = 1$ that the domain is an annulus whenever the best approximation to $\overline{z}$ is $\frac{c}{z}$. In the proof we use Lemma 3.9 which, via the argument principle, shows that the boundary of the domain consists of two boundary components. However, the hypothesis of the lemma assumes that the boundary of the domain is analytic, and that $\overline{z} \neq \frac{c}{z}$ on every boundary component of the domain. If $\overline{z} = \frac{c}{z}$ on some component, then that boundary contour is a circle but already our argument that the boundary has two components fails since the argument principle can only estimate the change in the argument of f^* on the remaining components. It should be possible to coach the proof of Theorem 3.10 to include the case when $\overline{z} = \frac{c}{z}$ on a boundary component and to extend it to all $p \geq 1$, but we have not been able to do it.

References

[1] Adams R. A., *Sobolev Spaces,* Academic Press, New York,1975.

[2] Ahlfors L., Beurling A., *Conformal invariants and function theoretic null sets*, Acta Math., 83 (1950) 101-129.

[3] Alexander H., *Projections of polynomial hulls*, J. Funct. Anal. **3** (1973), 13-19.

[4] Beneteau C., Khavinson D., *The isoperimetric inequality via approximation theory and free boundary problems*, Comput. Methods Funct. Theory, 6 (2006), No. 2, 253–274.

[5] Davis P., *The Schwarz Function and its Applications*, Carus Math. Monographs 17, Math. Assoc. of America, 1974.

[6] Duren P., *Theory of $\mathbb{H}^p$ Spaces*, Academic Press, New York, 1980.

[7] Duren P., Schuster, A., *Bergman Spaces*, American Mathematical Society, Providence, Rhode Island, 2004.

[8] Ebenfelt P., Khavinson D. and H. S., Shapiro, *A free boundary problem related to single-layer potentials,* Ann. Acad. Scie. Fenn. Math., **27** (2002), 21-46.

[9] Evans L., *Partial Differential Equations*, American mathematical Society, Providence, Rhode Island, 1998.

[10] Gamelin T., Khavinson D., *The isoperimetric inequality and rational approximation*, Amer. Math. Monthly **96** (1989), 18-30.

[11] Gustafsson B., Khavinson D., *Approximation by harmonic vector fields*, Houston J. Math. **20** (1994), no.1, 75-92.

[12] Khavinson D., *An isoperimetric problem*, Linear and Complex Analysis, Problem Book 3, Part II, V.P. Khavin and N.K. Nikolsky, eds., Lecture Notes Math., Springer-Verlag, **1574** (1994), 133-135.

[13] Khavinson D., *Annihilating measures of the algebra R(X)*, J. Funct. Anal. **58** (1984), 175-193.

[14] Khavinson D., *Smirnov classes of analytic functions in multiply connected domains,* Appendix to the English translation of *Foundations of the Theory of Extremal Problems for Bounded Analytic Functions and various generalizations of them,* by S. Ya. Khavinson, Amer. Math. Soc. Transl. (2), **129** (1986), 57-61.

[15] Khavinson D., *Symmetry and uniform approximation by analytic functions*, Proc. Amer. Math. Soc. **101** (1987), 475-483.

[16] Khavinson D., Luecking D., *On an extremal problem in the theory of rational approximation,* J. Approx. Th. **50** (1987), 127-132.

[17] Khavinson D., McCarthy J. and Shapiro H. S., *Best approximation in the mean by analytic and harmonic functions*, Indiana University Mathematics Journal, (4), **49** (2000), 1481-1513.

[18] Khavinson D., Stessin M., *Certain linear extremal problems in Bergman spaces of analytic functions*, Indiana Univ. Math J., **46** (1997), 933-974.

[19] Khavinson S. Ya., *Foundations of the Theory of Extremal Problems for Bounded Analytic Functions and various generalizations of them*, Amer. Math. Soc. Transl. (2), **129** (1986), 1-56.

[20] Shapiro H. S., *The Schwarz Function and its generalization to higher dimensions*, University of Arkansas Lecture Notes in Mathematical Sciences, **9** Wiley, (1992).

DEPARTMENT OF MATHEMATICS AND PHYSICS, ROCKHURST UNIVERSITY, KANSAS CITY, MISSOURI, 64110

E-mail address: guadarrama@rockhurst.edu

DEPARTMENT OF MATHEMATICS, UNIVERSITY OF SOUTH FLORIDA, TAMPA, FLORIDA 33620

E-mail address: dkhavins@cas.usf.edu

Contemporary Mathematics
Volume **454**, 2008

A General View of Multipliers and Composition Operators II

Don Hadwin and Eric Nordgren

1. Introduction

In this paper we continue our investigation of multiplier pairs begun in [**5**]. We give an alternative view of a multiplier pair in terms of an algebra of operators with a separating cyclic vector. A case of particular importance is $L^{\infty}[0,1]$, and we are led to study symmetric norms in this context and then unitarily invariant norms on type II_1 factor von Neumann algebras. In addition we start an investigation of multiplier pairs of tensor products.

2. Preliminaries

We call a pair (X,Y) a *multiplier pair* provided X is a Banach space, Y is a Hausdorff topological vector space, $X \subset Y$, and the inclusion map is continuous. Moreover, suppose we have a bilinear map (multiplication) from $m : X \times X \to Y$, with the notation $m(u,v) = u \cdot v$ such that

(1) m is separately continuous,
(2) The sets $\mathcal{L}_0 = \{x \in X : x \cdot X \subset X\}$ and $\mathcal{R}_0 = \{x \in X : X \cdot x \subset X\}$ are dense in X,
(3) There is an $e \in X$ such that, for every $x \in X$, $x \cdot e = e \cdot x = x$.
(4) There are dense subsets $E \subset \mathcal{L}_0, F \subset X, G \subset \mathcal{R}_0$ such that,
$$(u \cdot v) \cdot w = u \cdot (v \cdot w)$$
whenever $u \in E, v \in F, w \in G$.

If $x \in X$ we define L_x and R_x from X into Y by
$$L_x w = x \cdot w \text{ and } R_x w = w \cdot x,$$
where the domain of L_x is $Dom(L_x) = \{w \in X : x \cdot w \in X\}$ and the domain of R_x is $Dom(R_x) = \{w \in X : w \cdot x \in X\}$. We define $\mathcal{L} = \{L_x : x \in \mathcal{L}_0\}$ and $\mathcal{R} = \{R_x : x \in \mathcal{R}_0\}$.

THEOREM 2.1. *The following are true.*

(1) *The multiplication $\cdot$ is jointly continuous from $X \times X$ to Y.*

2000 *Mathematics Subject Classification.* Primary 46B28.

(2) *For every* $x \in X$, L_x, R_x *are densely defined closed operators on* X.
(3) L_x *is bounded on* $\mathcal{R}_0$ *if and only if* $x \in \mathcal{L}_0$, *and* R_x *is bounded on* $\mathcal{L}_0$ *if and only if* $x \in \mathcal{R}_0$.
(4) $\mathcal{L}, \mathcal{R} \subset B(X)$.
(5) *If* $u, v \in \mathcal{L}_0$ *or* $v, w \in \mathcal{R}_0$ *or* $u \in \mathcal{L}_0, w \in \mathcal{R}_0$, *then*

$$(u \cdot v) \cdot w = u \cdot (v \cdot w).$$

(6) $\mathcal{L}' = \mathcal{R}$ *and* $\mathcal{R}' = \mathcal{L}$.
(7) $L_v L_w = L_{v \cdot w}$ *if* $v, w \in \mathcal{L}_0$ *and* $R_v R_w = R_{w \cdot v}$ *if* $v, w \in \mathcal{R}_0$.

It was proved in [**5**] that it is possible to choose Y so that Y is a Banach space and $\operatorname{ball} Y$ is the closed convex hull $K = \operatorname{co}((\operatorname{ball} X) \cdot (\operatorname{ball} X))$ if and only if K contains no lines (i.e., $\mathbb{R}x$ for some nonzero vector x). In this case (X, Y) is called a *natural multiplier pair* and Y is called the *cospace* of X. A cospace exists whenever Y can be chosen so that the continuous linear functionals on Y separate the points of Y.

In the multiplier pair setting a notion of composition operator was defined that coincides with the usual notion in all of the classical cases. Suppose $\alpha : \mathcal{L} \to \mathcal{L}$ is a unital algebra homomorphism. Since $\mathcal{L}$ and $\mathcal{L}_0$ are isomorphic ($x \mapsto L_x$), α induces a unital algebra homomorphism $\hat{\alpha} : \mathcal{L}_0 \to \mathcal{L}_0$ defined by

$$L_{\hat{\alpha}(x)} = \alpha (L_x).$$

If $\hat{\alpha}$ is bounded (closable) on $\mathcal{L}_0$, we denote its continuous extension (closure) on X by C_α. We call C_α the (left) *composition operator induced by* α. Similarly, if $\beta : \mathcal{R} \to \mathcal{R}$ is a unital algebra homomorphism, we can define a unital algebra homomorphism $\hat{\beta} : \mathcal{R}_0 \to \mathcal{R}_0$ by

$$R_{\hat{\beta}(x)} = \beta (R_x).$$

Note that the multiplicativity of $\hat{\beta}$ follows from

$$R_{\hat{\beta}(x \cdot y)} = \beta (R_{x \cdot y}) = \beta (R_y R_x) = \beta (R_y) \beta (R_x) = R_{\hat{\beta}(y)} R_{\hat{\beta}(x)} = R_{\hat{\beta}(x) \cdot \hat{\beta}(y)}.$$

We denote the continuous extension (closure) of $\hat{\beta}$ on X by C^β, and we call C^β the (right) *composition operator induced by* β.

3. Completing Norms on $L^\infty [0,1]$

There is another way to view multiplier pairs. Suppose X is a Banach space, and $\mathcal{A}$ a unital weak-operator closed algebra of operators on X having a separating cyclic vector e. The identification of $\mathcal{A}$ with $\mathcal{A}e$ induces a multiplication $\cdot : \mathcal{A}e \times \mathcal{A}e \to \mathcal{A}e$ defined by

$$(Ae) \cdot (Be) = (AB) e.$$

We get a multiplier pair when we can extend this multiplication to all of X (allowing the product to be contained in a larger space Y. From this point of view, we can actually eliminate X. We can define a new norm ν on the Banach algebra $\mathcal{A}$ by

$$\nu (A) = \|Ae\|.$$

In this way, we can view X as the completion of $\mathcal{A}$ with respect to the norm ν. Note that we have

$$\nu (AB) \leq \|A\| \nu (B).$$

In this section we begin an investigation of norms on $L^{\infty}[0,1]$ so that the pointwise multiplication on $L^{\infty}[0,1]$ extends to a multiplication on the completion that gives rise to a multiplier pair. If $L^{\infty}[0,1]$ is contained in the multipliers on the completion of $L^{\infty}[0,1]$, then, with an equivalent norm [**6**], we have $\|fg\| \leq \|f\|_{\infty}\|g\|$ and $\|1\| = 1$ always hold. Let $X_{\|\|}$ denote the completion of $L^{\infty}[0,1]$ with respect to the norm $\|\|$.

Define $\lambda_{\|\|}, \eta_{\|\|} : [0,1] \to (0,\infty)$ by

$$\lambda_{\|\|}(t) = \inf\{\|\chi_E\| : \mu(E) = t\},$$
$$\eta_{\|\|}(t) = \sup\{\|\chi_E\| : \mu(E) = t\},$$

where μ is Lebesgue measure. We will restrict ourselves to norms for which the completions can be realized as measurable functions on $[0,1]$. Let Y be the topological vector space of all complex μ-measurable functions on $[0,1]$ with the topology of convergence in measure.

PROPOSITION 3.1. *Suppose $\|\|$ is a norm on $L^{\infty}[0,1]$ such that $\|1\| = 1$, $\|fg\| \leq \|f\|_{\infty}\|g\|$ and $\lambda_{\|\|}(t) > 0$ for every $t \in (0,1]$ and every $f,g \in L^{\infty}[0,1]$. Suppose also that $\lim_{t\to 0^+}\eta_{\|\|}(t) = 0$. Then*

(1) *$\|f\| \leq \|f\|_{\infty}$ for every $f \in L^{\infty}[0,1]$*
(2) *$\|f_n\| \to 0$ if and only if $\{f_n\}$ is $\|\|$-cauchy and $f_n \to 0$ in measure.*
(3) *$X_{\|\|} \subset Y$*
(4) *$(X_{\|\|}, Y)$ is a multiplier pair with pointwise (a.e.) multiplication.*
(5) *$\mathcal{L}_0 = \mathcal{R}_0 = L^{\infty}[0,1]$ and $\|L_f\| = \|f\|_{\infty}$ always holds.*

Proof. (1) This follows from $\|f\| = \|f\cdot 1\| \leq \|f\|_{\infty}\|1\| = \|f\|_{\infty}$.

(2) Suppose $\|f_n\| \to 0$. Clearly, $\{f_n\}$ is $\|\|$-cauchy. Also if $\varepsilon > 0$ and $E_n = \{x \in [0,1] : |f_n(x)| \geq \varepsilon\}$, then

$$\|f_n\| \geq \|f_n\chi_{E_n}\| \geq \varepsilon\|\chi_{E_n}\|.$$

Since $\lambda_{\|\|}(t) > 0$ for every $t \in (0,1]$ it follows that $\mu(E_n) \to 0$. Hence $f_n \to 0$ in measure.

Conversely, suppose $\{f_n\}$ is $\|\|$-cauchy and $f_n \to 0$ in measure, and $\|f_n\| \nrightarrow 0$. By taking a subsequence and normalizing, we can assume that $r \geq \|f_n\| \geq 1$ for every $n \in \mathbb{N}$ and some $r > 1$. Choose N so that $m,n \geq N$ implies $\|f_n - f_m\| < 1/3$. Let $E_m = \{x \in [0,1] : |f_m(x)| \geq 1/3\}$. Since $f_m \to 0$ in measure, $\mu(E_m) \to 0$. Since $\lim_{t\to 0^+}\eta_{\|\|}(t) = 0$, we have $\|\chi_{E_m}\| \to 0$. Then we have

$$\begin{aligned}1 \leq \|f_N\| &\leq \|(f_N - f_m)(1-\chi_{E_m})\| + \|\chi_{E_m}\|\|f_N\| + \|f_m(1-\chi_{E_m})\| \\ &\leq \|f_N - f_m\| + \|\chi_{E_m}\|\|f_N\|_{\infty} + \|f_m(1-\chi_E)\|_{\infty} \\ &\leq 2/3 + \|\chi_{E_m}\|\|f_N\|_{\infty} \to 2/3,\end{aligned}$$

which is a contradiction.

(3) It follows from (2) that the inclusion map from $L^{\infty}[0,1]$ into Y extends to a continuous injective map from $X_{\|\|}$ into Y.

(4) The continuity of the multiplication follows from (2), and the other properties are obvious.

(5) Suppose $f \in \mathcal{L}_0$, $r > 0$ and $E = \{x \in [0,1] : |f(x)| \geq r\}$. Then

$$\|L_f\chi_E\|\|\chi_E\| \geq \|L_f\chi_E\| \geq r\|\chi_E\|,$$

so if $\mu(E) > 0$, it follows that $\|L_f\| \geq r$. Hence $\|f\|_\infty \leq \|L_f\| < \infty$. Statement (1) implies that $\|L_f\| \leq \|f\|_\infty$. ∎

We say that a norm $\|\|$ on $L^\infty[0,1]$ (with respect to Lebesgue measure μ) is a *symmetric norm* if

(1) $\|1\| = 1$
(2) $\||f|\| = \|f\|$ for every $f \in L^\infty[0,1]$
(3) $\|f \circ \varphi\| = \|f\|$ for every $f \in L^\infty[0,1]$ and every invertible, measure-preserving $\varphi : [0,1] \to [0,1]$.

We say that a symmetric norm $\|\|$ on $L^\infty[0,1]$ is *continuous* if $\lim_{t\to 0^+} \|\chi_{[0,t)}\| = 0$.

Note that $\|\chi_E\| = \lambda(\mu(E)) > 0$ whenever $\|\|$ is a symmetric norm and E is a set of strictly positive measure.

THEOREM 3.2. *Suppose $\|\|$ is a symmetric norm on $L^\infty[0,1]$. Then*

(1) $\|fg\| \leq \|f\| \|g\|_\infty$ *for every* $f, g \in L^\infty[0,1]$
(2) $\|f\|_1 \leq \|f\| \leq \|f\|_\infty$ *for every* $f \in L^\infty[0,1]$
(3) $\|\|$ *is equivalent to* $\|\|_\infty$ *if and only if* $\|\|$ *is not continuous.*
(4) *If* $\{f_n\}$ *is a* $\|\|$*-Cauchy sequence in* $L^\infty[0,1]$ *and* $f_n \to 0$ *in measure, then* $\|f_n\| \to 0$.
(5) *The (pointwise) multiplication extends uniquely to a bilinear map from* $X_{\|\|} \times X_{\|\|}$ *to* Y *that makes* $(X_{\|\|}, Y)$ *into a multiplier pair.*
(6) $X_{\|\|}$ *has a cospace only if*

$$\limsup_{t\to 0^+} \frac{t}{\|\chi_{[0,t]}\|^2} < \infty.$$

(7) *If* $X_{\|\|}$ *has a cospace, the cospace norm (normalized so* 1 *has norm* 1*) is a symmetric norm.*
(8) *If* $\|\|$ *is continuous, then on* $\{f \in L^\infty[0,1] : \|f\|_\infty \leq 1\}$ *the* $\|\|$*-topology coincides with the topology of convergence in measure.*
(9) *In the multiplier pair* $(X_{\|\|}, Y)$ *we have* $\mathcal{L}_0 = \mathcal{R}_0 = L^\infty[0,1]$ *and, for every* $f \in L^\infty[0,1]$,

$$\|L_f\| = \|R_f\| = \|f\|_\infty.$$

Proof. (1) It follows from $\|f\| = \||f|\|$ that multiplication by a function g with $|g| = 1$ is an isometry on $L^\infty[0,1]$ with respect to $\|\|$. However, every $h \in L^\infty[0,1]$ with $\|h\|_\infty \leq 1$ can be written as the average of two functions g_1 and g_2 with $|g_1| = |g_2| = 1$. Thus (1) is true.

(2) It follows from (1) and $\|1\| = 1$ that $\|f\| \leq \|f\|_\infty$ for every $f \in L^\infty[0,1]$. For the other inequality, suppose $f = \sum_{k=1}^m \alpha_k \chi_{E_k}$ is a simple function with $\{E_1, \ldots, E_m\}$ a measurable partition of $[0,1]$. Let $[x]$ denote the greatest integer function of x. let $n \in \mathbb{N}$, and let $s_k = [n\mu(E_k)]$ for $1 \leq k \leq m$. Choose a measurable partition $\{F_1, \ldots, F_n\}$ of $[0,1]$ so that each F_j has measure $\frac{1}{n}$ and, for each k, $1 \leq k \leq m$, E_k contains exactly s_k of the F_j's. Next choose a measure-preserving isomorpism $\varphi : [0,1] \to [0,1]$ so that $\varphi(F_1) = F_2, \ldots, \varphi(F_{n-1}) = F_n$, and $\varphi(F_n) = F_1$. Let $g_n = \frac{1}{n}\sum_{j=0}^{n-1} |f| \circ \varphi^j$. We then have

$$\|g_n\| \leq \frac{1}{n}\sum_{j=0}^{n-1} \||f| \circ \varphi^j\| = \|f\|.$$

However,

$$\begin{aligned}\left\| g_n - \int |f|\, d\mu \right\| &\leq \left\| g_n - \int |f|\, d\mu \right\|_\infty \\ &\leq \left| \sum_{k=1}^{m} |\alpha_k| \left(\mu(E_k) - \frac{[n\mu(E_k)]}{n} \right) \right| + \frac{m}{n} \|f\|_\infty \\ &\to 0\end{aligned}$$

as $n \to \infty$. Hence, $\|f\|_1 \leq \|f\|$ when f is a simple function. However, the inequality $\|f\| \leq \|f\|_\infty$ and the $\|\|_\infty$-density of the simple functions in $L^\infty[0,1]$ yields $\|f\|_1 \leq \|f\|$ for every $f \in L^\infty[0,1]$.

(3) The only if part is obvious. Suppose $\lim_{t\to 0^+} \|\chi_{[0,t]}\| = L > 0$. It follows from invariance that $\|\chi_E\| \geq L$ whenever $\mu(E) > 0$. If $r < \|f\|_\infty$, then $E = \{x : |f(x)| \geq r\}$ has positive measure and

$$\|f\| \geq \|\chi_E |f|\| \geq r \|\chi_E\| \geq rL.$$

It follows that $\|f\| \geq L \|f\|_\infty$.

(4) If $\|\|$ is equivalent to $\|\|_\infty$ the assertion is obvious. Hence we can assume that $\|\|$ is continuous. In this case we can apply part (2) of Proposition 3.1.

(5) If $f \in L^\infty[0,1]$ and $\varepsilon > 0$ and $E = \{x \in [0,1] : |f_m(x)| \geq \varepsilon\}$, then

$$\|f\| \geq \|f\chi_E\| \geq \varepsilon \|\chi_E\| \geq \varepsilon \|\chi_E\|_1 = \varepsilon\mu(E).$$

Hence the inclusion map from $L^\infty[0,1]$ to Y is continuous with $\|\|$ on $L^\infty[0,1]$ and convergence in measure on Y. By (4), the inclusion map extends to a continuous injective linear map from $X_{\|\|}$ into Y. Thus the statement (5) is clear.

(6) Suppose $\limsup_{t\to 0^+} \frac{t}{\|\chi_{[0,t)}\|^2} = \infty$. Clearly, $\lim_{t\to 0^+} \|\chi_{[0,t)}\| = 0$. Choose a decreasing sequence $\{t_n\}$ in $(0,1/4]$ with $t_n \to 0$ such that $\frac{t_n}{\|\chi_{[0,t_n)}\|^2} \to \infty$. Let $k_n = \left[\frac{1}{t_n}\right]$ Then

$$\frac{1}{k_n} \sum_{j=0}^{k_n - 1} \frac{1}{\|\chi_{[0,t_n)}\|^2} \left(\chi_{[jt_n,(j+1)t_n)} \chi_{[0,\frac{1}{2})} \right) \in \mathrm{co}\left(\mathrm{ball}\left(X_{\|\|}\right) \cdot \mathrm{ball}\left(X_{\|\|}\right)\right).$$

However,

$$\frac{1}{k_n} \sum_{j=0}^{k_n - 1} \frac{1}{\|\chi_{[0,t_n)}\|^2} \left(\chi_{[jt_n,(j+1)t_n)} \chi_{[0,\frac{1}{2})} \right) = \frac{1}{t_n k_n} \frac{t_n}{\|\chi_{[0,t_n)}\|^2} \chi_{[0,\frac{1}{2})},$$

and, since $t_n k_n \to 1$, we see that $\mathrm{co}\left(\mathrm{ball}\left(X_{\|\|}\right) \cdot \mathrm{ball}\left(X_{\|\|}\right)\right)$ contains the line $\mathbb{R}\chi_{[0,\frac{1}{2})}$. Hence there is no cospace.

(7) Let $K = \mathrm{co}\left(\mathrm{ball}\left(X_{\|\|}\right) \cdot \mathrm{ball}\left(X_{\|\|}\right)\right)$. Clearly K is convex absorbing and balaced in $L^\infty[0,1]$ and the existence of a cospace says that K contains no line, so that the Minkowski functional $\|\|_K$ is a norm on $L^\infty[0,1]$. It is clear from the fact that $\|\|$ is a symmetric norm that $f \in K$ if and only if $|f| \in K$ if and only if $f \circ \varphi \in K$ for every measure-preserving Borel isomorphism $\varphi : [0,1] \to [0,1]$. It follows that $\|f\|_K = \||f|\|_K = \|f \circ \varphi\|_K$ for every $f \in L^\infty[0,1]$ and every measure-preserving Borel isomorphism φ on $[0,1]$. Moreover, we have $\|fg\|_K \leq \|f\| \|g\|$ for all $f, g \in L^\infty[0,1]$. Hence $\|1\|_K \leq 1$.

(8) The fact that $\|\|_1 \leq \|\|\|$ implies that convergence in $\|\|\|$ implies convergence in measure. Suppose $\|\|\|$ is continuous and $\{f_n\}$ is in the unit ball $L^\infty[0,1]$ and $f_n \to f$ in measure. Suppose $\varepsilon > 0$. The continuity of $\|\|\|$allows us to choose $\delta > 0$ such that $\mu(E) < \delta$ implies $\|\chi_E\| < \varepsilon/4$, and convergence in measure implies that there is an n_n such that $\mu(\{x : |f(x) - f_n(x)| \geq \varepsilon/2\}) < \delta$ whenever $n \geq n_0$. It follows that, for $n \geq n_0$,

$$\|f - f_n\| < \frac{\varepsilon}{2} + \frac{\varepsilon}{4}\|f - f_n\|_\infty \leq \varepsilon.$$

(9) Suppose $f \in \mathcal{L}_0$. Then L_f is bounded on $X_{\|\|\|}$ so

$$\|fg\| \leq \|L_f\| \|g\|$$

for every $g \in L^\infty[0,1]$. If $\|f\|_\infty > \|L_f\|$, then the essential range of f contains a number λ with $|\lambda| > \|L_f\|$. For every $\varepsilon > 0$ the set $E(\varepsilon) = \{x : |f(x) - \lambda| < \varepsilon\}$ has positive measure, so if $g_\varepsilon = \chi_{E(\varepsilon)}/\left\|\chi_{E(\varepsilon)}\right\|$, then

$$\|(L_f - \lambda) g_\varepsilon\| \leq \left\|(f - \lambda)\chi_{E(\varepsilon)}\right\|_\infty \|g_\varepsilon\| \leq \varepsilon.$$

Hence $\lambda \in \sigma(L_f)$, which contradicts $|\lambda| > \|L_f\|$. Thus $f \in L^\infty[0,1]$ and $\|f\|_\infty \leq \|L_f\|$. However, it follows from (1) that $\|L_f\| \leq \|f\|_\infty$. ∎

4. Unitarily Invariant Norms on a Finite Factor

Suppose $\mathcal{M}$ is a II_1 factor von Neumann algebra with a faithful normal trace τ. A norm ν on $\mathcal{M}$ is a *unitarily invariant norm* if $\nu(1) = 1$ and $\nu(UTV) = \nu(T)$ for every $T \in \mathcal{M}$ and all unitaries $U, V \in \mathcal{M}$. The Russo-Dye theorem tells us that the closed unit ball of $\mathcal{M}$ is the norm-closed convex hull of the set of unitaries in $\mathcal{M}$, so we have $\nu(T) \leq \|T\|$ for every $T \in \mathcal{M}$.

Since every T in $\mathcal{M}$ has a polar decomposition $T = U(T^*T)^{\frac{1}{2}}$ with U unitary, it follows that $\nu(T) = \nu\left((T^*T)^{\frac{1}{2}}\right)$ for every $T \in \mathcal{M}$. If we expand the set of spectral projections $\chi_E\left((T^*T)^{\frac{1}{2}}\right)$, with E ranging over intervals of the form $[0,s)$ or $[0,s]$, to a maximal chain of projections $\{P_t : t \in [0,1]\}$ with each $\tau(P_t) = t$, we see that $(T^*T)^{\frac{1}{2}} \in \{P_t : t \in [0,1]\}''$. Moreover, $\{P_t : t \in [0,1]\}''$ is tracially isomorphic to $L^\infty[0,1]$. More precisely, the map $\chi_{[0,t)} \to P_t$ extends uniquely to a $*$-isomorphism $\pi : L^\infty[0,1] \to \{P_t : t \in [0,1]\}''$ such that, for every $f \in L^\infty[0,1]$,

$$\tau(\pi(f)) = \int_{[0,1]} f d\mu.$$

It follows from results of Huiru Ding and the first author [**2**] that if $\pi_1, \pi_2 : L^\infty[0,1] \to \mathcal{M}$ are unital $*$-homomorphisms such that $\tau \circ \pi_1 = \tau \circ \pi_2$, then there is net $\{U_\lambda\}$ of unitaries in $\mathcal{M}$ such that, for every $f \in L^\infty[0,1]$,

$$\nu(U_\lambda^* \pi_1(f) U_\lambda - \pi_2(f)) \leq \|U_\lambda^* \pi_1(f) U_\lambda - \pi_2(f)\| \to 0,$$

so $\nu \circ \pi_1 = \nu \circ \pi_2$. Hence the norm $\nu \circ \pi$ is independent of the element $T \in \mathcal{M}$ or the representation π. Moreover, if $\alpha : [0,1] \to [0,1]$ is a bijective measurable measure-preserving transformation, then, for every $f \in L^\infty[0,1]$,

$$\tau(\pi(f \circ \alpha)) = \int_{[0,1]} (f \circ \alpha) d\mu = \int_{[0,1]} f d\mu.$$

Hence the norm $\nu \circ \pi$ is a symmetric norm on $L^{\infty}[0,1]$.

It follows from results in [**3**] and [**4**] that every symmetric norm on $L^{\infty}[0,1]$ corresponds to a unitarily invariant norm on $\mathcal{M}$.

There is also a notion of *convergence in measure* introduced by Nelson [**10**]. A net $\{T_\lambda\}$ in $\mathcal{M}$ converges *in measure* to $T \in \mathcal{M}$ if and only if, for every $\varepsilon > 0$ there is a projection P with $\tau(P) < \varepsilon$ and there is a λ_0 such that

$$\|(T_\lambda - T) - P(T_\lambda - T)P\| < \varepsilon$$

whenever $\lambda \geq \lambda_0$. Let $\mathcal{M}^\tau$ denote the completion of $\mathcal{M}$ with respect to the topology of convergence in measure. It is a fact [**10**] that the multiplication on $\mathcal{M}$ extends to a multiplication on $\mathcal{M}^\tau$ that is jointly continuous.

One class of examples of unitarily invariant norms are the p-norms, $\|\|_p$, for $1 \leq p < \infty$ defined by

$$\|T\|_p = \tau\left((T^*T)^{p/2}\right)^{\frac{1}{p}}.$$

The norm $\|\|_2$ arises naturally in the GNS construction. In [**5**] we showed that the completion X of $\mathcal{M}$ with respect to $\|\|_1$ is a subset of Y and that $(X, \mathcal{M}^\tau)$ is a multiplier pair with $\mathcal{L}_0 = \mathcal{R}_0 = \mathcal{M}$. Here we show that this result extends to all unitarily invariant norms on $\mathcal{M}$. If ν is a unitarily invariant norm on $\mathcal{M}$, let $\mathcal{M}_\nu$ denote the completion with respect to ν. We call a unitarily invariant norm ν *continuous* if the corresponding symmetric norm on $L^{\infty}[0,1]$ is continuous, equivalently, if

$$\lim_{\tau(P)\to 0} \|P\| = 0$$

as P ranges over the projections in $\mathcal{M}$.

THEOREM 4.1. *Suppose ν is a unitarily invariant norm on a II_1 factor $\mathcal{M}$. Then*

(1) $\nu(ST) \leq \nu(S)\|T\|$ *and* $\nu(ST) \leq \nu(T)\|S\|$ *for every* $S, T \in \mathcal{M}$
(2) $\|T\|_1 \leq \nu(T) \leq \|T\|$ *for every* $T \in \mathcal{M}$
(3) $\|\|$ *is equivalent to* $\|\|_\infty$ *if and only if* $\lim_{t\to 0^+} \left\|\chi_{[0,t]}\right\| > 0$
(4) *The (pointwise) multiplication extends uniquely to a bilinear map from* $\mathcal{M}_\nu \times \mathcal{M}_\nu$ *to* $\mathcal{M}^\tau$ *that makes* $(\mathcal{M}_\nu, \mathcal{M}^\tau)$ *into a multiplier pair.*
(5) $\mathcal{M}_\nu$ *has a cospace only if*

$$\limsup_{P=P^2=P^*,\ \tau(P)\to 0^+} \frac{\tau(P)}{\|P\|^2} < \infty.$$

(6) *If* $\mathcal{M}_\nu$ *has a cospace, the cospace norm (normalized at 1) is a unitarily-invariant norm on* $\mathcal{M}$.
(7) *If* ν *is continuous, then on* ball $\mathcal{M}$ *the* ν*-topology, the* $*$*-strong operator topology and the topology of convergence in measure coincide.*
(8) *In the multiplier pair* $(\mathcal{M}_\nu, \mathcal{M}^\tau)$ *we have* $\mathcal{L}_0 = \mathcal{R}_0 = \mathcal{M}$ *and, for every* $A \in \mathcal{M}$, *we have* $\|L_A\| = \|R_A\| = \|A\|$.

Proof. (1) We already proved this above.

(2) This follows from part (2) of Theorem 3.2 and the description of $\nu \circ \pi$ above.

(4) This is the result of Nelson mentioned above.

(3), (5), (6), (7) These follow from their analogues (or their proofs) in Theorem 3.2.

(8) It follows from (1) that $\|L_T\|, \|R_T\| \leq \|T\|$ for every $T \in \mathcal{M}$ and it follows from [**6**] that $\|L_T\| = \|R_T\| = \|T\|$ for every $T \in \mathcal{M}$. If ν is equivalent to $\|\|$ on $\mathcal{M}$, the desired assertion is obvious. Hence we can assume that ν is continuous. Suppose $T \in \mathcal{L}_0$. It follows from $\|A\|_1 \leq \nu(A)$ and the definition of convergence in measure that, for every $n \in \mathbb{N}$ there is a projection $P_n \in \mathcal{M}$ with $\tau(1 - P_n) < 1/n$ with $TP_n \in \mathcal{M}$. It is clear that $\|TP_n\| = \|L_{TP_n}\| \leq \|L_T\|$. Since ν is continuous $\nu(1 - P_n) \to 0$, and since L_T is continuous, $\nu(T(1 - P_n)) \to 0$. Hence TP_n converges to T in ν-norm, but it follows from (7) that $T \in \mathcal{M}$ and $\|T\| \leq \|L_T\|$. ∎

REMARK 4.2. In an early version of [**7**] the proof of proposition 7.5 included a proof that if $\{a_n\}$ is a norm-bounded sequence in $\mathcal{M}$ and $\|a_n\|_p \to 0$ for some $0 < p < 1$, then $\|a_n\|_2 \to 0$. However, $\|a_n\|_p \to 0$ easily implies $a_n \to 0$ in measure, and by part (7) of the preceding theorem that $\nu(a_n) \to 0$ for any continuous unitarily invariant norm ν on $\mathcal{M}$.

5. Tensor Products

We now want to consider tensor products of multiplier pairs. Suppose (X_i, Y_i) are multiplier pairs for $i = 1, 2$. Let $X = X_1 \otimes X_2$ and $Y = Y_1 \otimes Y_2$ be the algebraic tensor products. We can define a multiplication $\cdot$ on X with values in Y by

$$(a_1 \otimes b_1) \cdot (a_2 \otimes b_2) = (a_1 \cdot a_2) \otimes (b_1 \cdot b_2).$$

The first question that interests us is, if $\|\|_X$ is a tensor norm on X, i.e.,

$$\|a \otimes b\|_X = \|a\| \|b\|,$$

then can we extend the multiplication on X to the completion X^- of X that creates a multiplier pair? An important special case is when (X_i, Y_i) are natural multiplier pairs and $\|\|_X$ is a tensor norm on X, when does there exist a tensor norm $\|\|_Y$ on Y so that the multiplication extends so that (X^-, Y^-) is a natural multiplier pair? One way to show that the latter question has a negative answer is to show that

$$K = \text{co}((\text{ball}\, X) \cdot (\text{ball}\, X))$$

contains a line, although this does not answer the first question. If K does not contain any lines, then we have

$$\|u \cdot v\|_Y \leq \|u\|_X \|v\|_X$$

for all $u, v \in X$. The only difficulty is that, although the inclusion map from X to Y is injective and satisfies

$$\|x\|_Y \leq \|1 \otimes 1\|_X \|x\|_X,$$

it is not clear that the inclusion from X^- to Y^- is still injective.

Of course all of these problems go away when X_1 and X_2 are finite-dimensional, and since $\mathbb{C}^m \otimes \mathbb{C}^n$ is the space $\mathcal{M}_{m\times n}(\mathbb{C})$ of $m\times n$ matrices, different multiplications on $\mathbb{C}^m$ and $\mathbb{C}^n$ yield multiplications on $\mathcal{M}_{m\times n}(\mathbb{C})$; in particular, coordinatewise multiplication on $\mathbb{C}^m$ and $\mathbb{C}^n$ yields the Shur (entrywise) product on $\mathcal{M}_{m\times n}(\mathbb{C})$. These ideas lead to a myriad of finite-dimensional examples.

Of all the tensor norms on X there is a unique smallest one $\|\|_{\min}$ and a unique largest one $\|\|_{\max}$. Hence, if $\|\|_X$ is a tensor norm on X, then $\text{ball}_{\|\|_{\max}} X \subset \text{ball}_{\|\|_X} X \subset \text{ball}_{\|\|_{\min}} X$, so if $\text{co}\left(\left(\text{ball}_{\|\|_{\min}} X\right) \cdot \left(\text{ball}_{\|\|_{\min}} X\right)\right)$ contains no lines, then $\mathrm{co}\left(\left(\mathrm{ball}_{\|\|_X} X\right)\cdot\left(\mathrm{ball}_{\|\|_X} X\right)\right)$ contains no lines for every tensor norm $\|\|_X$. Hence the $\|\|_{\min}$ is a natural first case to check. Similarly, if

$$\mathrm{co}\left(\left(\mathrm{ball}_{\|\|_X} X\right)\cdot\left(\mathrm{ball}_{\|\|_X} X\right)\right)$$

contains no lines for some tensor norm $\|\|_X$, then $\mathrm{co}\left(\left(\mathrm{ball}_{\|\|_{\max}} X\right)\cdot\left(\mathrm{ball}_{\|\|_{\max}} X\right)\right)$ contains no lines. Here is an interesting example.

PROPOSITION 5.1. *If $X_1 = X_2 = L^2[0,1]$ with pointwise (a.e.) multiplication, then $\mathrm{co}\left(\left(\mathrm{ball}_{\|\|_{\min}} X\right)\cdot\left(\mathrm{ball}_{\|\|_{\min}} X\right)\right)$ contains a line.*

Proof. Note that $X_1 \otimes_{\min} X_2$ is the set of compact operators on $L^2[0,1]$ with the operator norm. Suppose $n \in \mathbb{N}$ and let $F_k = \sqrt{n}\chi_{\left[\frac{k}{n},\frac{k+1}{n}\right]}$ for $0 \le k < n$. Then $\|F_k\| = 1$ for each k and $\|F_i \otimes F_j\|_{\min} = 1$ for $1 \le i,j \le n$. For each integer s with $0 \le s < n$ let

$$T_s = \sum_{|i-j|=s \mod n} F_i \otimes F_j.$$

Since $\|\|_{\min}$is the operator norm, $\|T_s\|_{\min} = 1$. Moreover,

$$T_s \cdot T_s = \sum_{|i-j|=s \mod n} (F_i \cdot F_i) \otimes (F_j \cdot F_j) = n^2 \sum_{|i-j|=s \mod n} \chi_{\left[\frac{i}{n},\frac{i+1}{n}\right]} \otimes \chi_{\left[\frac{j}{n},\frac{j+1}{n}\right]}$$

is in $\left(\mathrm{ball}_{\|\|_{\min}} X\right)\cdot\left(\mathrm{ball}_{\|\|_{\min}} X\right)$ and

$$n(1 \otimes 1) = \frac{1}{n}\sum_{s=0}^{n-1} T_s \cdot T_s$$

is in $\mathrm{co}\left(\left(\mathrm{ball}_{\|\|_{\min}} X\right)\cdot\left(\mathrm{ball}_{\|\|_{\min}} X\right)\right)$. Since $\mathrm{co}\left(\left(\mathrm{ball}_{\|\|_{\min}} X\right)\cdot\left(\mathrm{ball}_{\|\|_{\min}} X\right)\right)$ contains 0 and is absolutely convex, it follows that $\mathrm{co}\left(\left(\mathrm{ball}_{\|\|_{\min}} X\right)\cdot\left(\mathrm{ball}_{\|\|_{\min}} X\right)\right)$ contains the line $\mathbb{R}(1 \otimes 1)$. ■

COROLLARY 5.2. *Suppose G is a countable discrete abelian group and $X_1 = X_2 = \ell^2(G)$ with the convolution product. Then $X_1 \otimes_{\min} X_2$ has a cospace if and only if G is finite.*

Proof. If $\hat{G}$ is the dual group of G, with Haar measure μ, then X_1 and X_2 are isometrically isomorphic to $L^2\left(\hat{G},\mu\right)$ with pointwise (a.e.) multiplication. If G is infinite, then μ is continuous and we are in the situation of the preceding theorem. ■

For the projective tensor product things work better. This has to do with the definition of $\|\|_{\max}$ on $V \otimes W$, i.e.,

$$\|s\| = \inf\left\{\sum_{k=1}^{n} \|v_k\|\,\|w_k\| : s = \sum_{k=1}^{n} v_k \otimes w_k\right\}.$$

If we let $t_k = \|v_k\|\,\|w_k\|$ and replace $v_k \otimes w_k$ with $\frac{v_k}{\|v_k\|} \otimes \frac{w_k}{\|w_k\|}$, we see that

$$\mathrm{ball}_{\|\|_{\max}}(V \otimes W) = \mathrm{co}\{v \otimes w : \|v_k\| = \|w_k\| = 1\}.$$

The following lemma, which is obvious from the definition of a cospace, shows the relationship between multiplier pairs and projective tensor products. In particular, it gives a representation of the cospace when it exists.

Lemma 5.3. *Suppose (X, Y) is a multiplier pair and $T : X \otimes X \to Y$ is the linear mapping defined by*

$$T\left(a \otimes b\right) = a \cdot b.$$

Then

(1) $T\left(\mathrm{ball}_{\|\|_{\max}}\left(X \otimes X\right)\right) = \mathrm{co}\left(\left(\mathrm{ball}\, X\right) \cdot \left(\mathrm{ball}\, X\right)\right)$

(2) *If Y is a cospace for X and $\hat{T} : X \otimes_{\max} X \to Y$ is the bounded extension of T, then the induced map from $\left(X \otimes_{\max} X\right) / \ker \hat{T}$ to Y is an isometric isomorphism.*

Suppose V_j and W_j are Banach spaces and $T_j : X_j \to Y_j$ is an injective bounded operator for $j = 1, 2$. Then the operator $T_1 \otimes_{\max} T_2 : X_1 \otimes_{\max} X_2 \to Y_1 \otimes_{\max} Y_2$ defined by $\left(T_1 \otimes_{\max} T_2\right)\left(v_1 \otimes v_2\right) = T_1 v_1 \otimes T_2 v_2$ may not be injective [**1**, p. 49]. However, if either V_1 or V_2 has the approximation property, then $T_1 \otimes_{\max} T_2$ must be injective.

Theorem 5.4. *Suppose (X_1, Y_1) and (X_2, Y_2) are natural multiplier pairs, with $\tau_i : X_i \to Y_i$ the inclusion maps. If $\tau_1 \otimes_{\max} \tau_2 : X_1 \otimes_{\max} X_2 \to Y_1 \otimes_{\max} Y_2$ is injective, then $\left(X_1 \otimes_{\max} X_2, Y_1 \otimes_{\max} Y_2\right)$ is a natural multiplier pair.*

Proof. The closed unit ball of Y_j is the closed convex hull of

$$\left\{a \cdot b : a, b \in X_j, \|a\| = \|b\| = 1\right\}$$

and the closed unit ball of $Y_1 \otimes Y_2$ is the closed convex hull of

$$\left\{y_1 \otimes y_2 : y_1 \in Y_1, y_2 \in Y_2, \|y_1\| = \|y_2\| = 1\right\},$$

so the closed unit ball of $Y_1 \otimes_{\max} Y_2$ is the closed convex hull of

$$\begin{aligned} E &= \left\{\left(a_1 \cdot a_2\right) \otimes \left(b_1 \cdot b_2\right) : a_1, a_2 \in X_1, a_2, b_2 \in X_2, \|a_j\| = \|b_j\| = 1\right\} \\ &= \left\{\left(a_1 \otimes b_1\right) \cdot \left(a_2 \otimes b_2\right) : a_1, a_2 \in X_1, a_2, b_2 \in X_2, \|a_j\| = \|b_j\| = 1\right\}. \end{aligned}$$

On the other hand the closed unit ball of $X = X_1 \otimes_{\max} X_2$ is the closed convex hull of

$$\left\{a_1 \cdot a_2 : a_1 \in X_1, a_2 \in X_2, \|a_1\| = \|a_2\| = 1\right\},$$

so the closed convex hull of $\left(\mathrm{ball}\, X\right) \cdot \left(\mathrm{ball}\, X\right)$ is also the closed convex hull of E defined above. ∎

Corollary 5.5. *If (X_1, Y_1) and (X_2, Y_2) are natural multiplier pairs, and if either X_1 or X_2 has the approximation property, then $\left(X_1 \otimes_{\max} X_2, Y_1 \otimes_{\max} Y_2\right)$ is a natural multiplier pair.*

Another class of tensor product examples comes from complex analysis. The Hardy space H^2 consists of all analytic functions in the open unit disc $\mathbb{D}$ with power series having square summable Taylor coefficients. The product of H^2 functions in the unit ball is in the unit ball of H^1, and every H^1 function in the unit ball can be factored into a product of two functions in the unit ball of H^2. As was pointed out in [**5**], H^1 is the cospace of H^2.

If the tensor product $H^2 \otimes H^2$ is given the Hilbert space norm, then its completion $H^2 \hat{\otimes}_2 H^2$ may be identified with $H^2(\mathbb{D}^2)$. The multiplication of elements of the tensor product defined by $(f_1 \otimes g_1) \cdot (f_2 \otimes g_2) = (f_1 f_2) \otimes (g_1 g_2)$ corresponds under this identification to ordinary multiplication of functions, and thus the cospace is identified with a subspace of $H^1(\mathbb{D}^2)$ but the situation is more complex than that of the disk. A result of Rosay [**12**] (see also Rudin [**12**] and Miles [**9**]) shows that functions in ball $H^1(\mathbb{D}^2)$ can not always be factored as products of functions in ball $H^2(\mathbb{D}^2)$. However a result of I. J. Lin [**8**] asserts that every element f in $H^1(\mathbb{D}^2)$ has a representation $f = \sum_{n=1}^{\infty} F_n G_n$ with $F_n, G_n \in H^2(\mathbb{D}^2)$ and $\sum_{n=1}^{\infty} \|F_n\| \|G_n\| \leq c \|f\|_1$. Since $\|f\|_1 \leq \sum_{n=1}^{\infty} \|F_n\| \|G_n\|$, the cospace is $H^1(\mathbb{D}^2)$ and the cospace norm is equivalent to the $H^1(\mathbb{D}^2)$ norm.

References

[1] Defant, Andreas and Floret, Klaus, *Tensor norms and ideals*, North-Holland, Amsterdam, 1993. MR1209438 (94e:46130)

[2] Ding, Huiru and Hadwin, Don, Approximate equivalence in von Neumann algebras, Sci. China Ser. A 48 (2005), no. 2, 239–247. MR2158602 (2006c:46050)

[3] Fack, Thierry Sur la notion de valeur caractéristique, J. Operator Theory 7 (1982), no. 2, 307–333. MR0658616 (84m:47012)

[4] Fack, Thierry and Kosaki, Hideki, Generalized s-numbers of τ-measurable operators, Pacific J. Math. 123 (1986), no. 2, 269–300. MR0840845 (87h:46122)

[5] Hadwin, D. and Nordgren, E., A general view of multipliers and composition operators, *Linear Algebra Appl.* **383** (2004), 187–211. MR2073904 (2005g:47123)

[6] Hadwin, Don and Orhon, Mehmet, A noncommutative theory of Bade functionals, Glasgow Math. J. 33 (1991), no. 1, 73–81. MR1089956 (91m:46085)

[7] Haagerup, U. and Schultz, H., Invariant subspaces for operators in a general II_1-factor, preprint.

[8] Lin, Ing Jer and Russo, Bernard, Applications of factorization in the Hardy spaces of the polydisk, *Interaction between functional analysis, harmonic analysis, and probability (Columbia, MO, 1994), 331–349, Lecture Notes in Pure and Appl. Math., 175, Dekker, New York*, 1996. MR1358170 (96g:46018)

[9] Miles, Joseph, A factorization theorem in $H^1(U^3)$, *Proc. Amer. Math. Soc.***52** (1975), 319–322. MR0374459 (51 #10659)

[10] Nelson, Notes on non-commutative integration, J. Functional Analysis, **15** (1974), 103–116. MR0355628 (50 #8102)

[11] Rosay, Jean-Pierre, Sur la non-factorisation des éléments de l'espace de Hardy $H^1(U^2)$, *Illinois J. Math.* **19**(1975), 479–482. MR0377098 (51 #13272)

[12] Rudin, Walter, *Function theory in polydiscs*, Benjamin, New York, 1969. MR0255841 (41 #501)

Department of Mathematics and Statistics, University of New Hampshire, Durham, NH 03824

E-mail address, D. Hadwin: don@math.unh.edu

E-mail address, E. Nordgren: ean@math.unh.edu

Contemporary Mathematics
Volume **454**, 2008

A General View of BMO and VMO

Don Hadwin and Hassan Yousefi

Dedicated to the memory of Paul Halmos

ABSTRACT. We formulate a very general setting in which the spaces BMO and VMO can be defined, and we prove several results in this general setting. We prove general versions of the John-Nirenberg theorem and characterizations of VMO. One main result is that VMO is never complemented in BMO.

1. Introduction

In this paper we construct a general setting in which functions of bounded mean oscillation BMO and vanishing mean oscillation VMO can be studied. One benefit of developing this general theory is that, when we throw away all unnecessary structure, we are forced to identify the "real reasons" that a theorem is true. For example, we prove a version of Sarason's characterization of VMO as the closure of the uniformly continuous functions in BMO that uses very simple ideas and doesn't depend on the clever, but specialized, use of a convolution argument. Another benefit is that proving a theorem in the general setting gives a proof in all the different realizations. For example we prove that VMO is not complemented in BMO, and one realization yields the conclusion that if μ is a continuous probability measure on a compact metric space X, then $C(X)$ is not complemented in $L^{\infty}(\mu)$. Another benefit is that the general setting provides a framework in which certain questions and concepts (e.g., products, homomorphisms) become natural that would not arise in a specific instance. In this setting we also prove a version of the John-Nirenberg theorem.

The notion of functions of *bounded mean oscillation* (BMO) made its first appearance in [**6**] where F. John studied BMO on $\mathbb{R}^n$ with Lebesgue measure. The celebrated *John-Nirenberg* inequality was proved in the next article in the same issue of Comm. Pure Appl. Math.[**7**]. Later a great deal of work was done on BMO on the circle with Haar measure.

2000 *Mathematics Subject Classification.* Primary 32A37, 46E30; Secondary 46B20.

Key words and phrases. BMO, VMO, John-Nirenberg Theorem, BMO triple, complemented subspace.

This paper is in final form and no version of it will be submitted for publication elsewhere.

Roughly speaking, functions of Bounded Mean Oscillation are the ones that on average, are not too far from the local averages of the function. To be more precise, an integrable function f defined on $\mathbb{R}^n$ is said to be in BMO if

$$\|f\|_* = \sup I(|f - I(f)|) < \infty,$$

where the supremum is taken over all cubes I that are Cartesian products of subintervals of the coordinate axes, $|I|$ is Lebesgue measure of the cube I, and $I(f) = \frac{1}{|I|}\int_I f$ is the average of the function on I. For BMO on the circle, we let I range over all arcs and let $|I|$ denote the Haar measure (normalized arc length) of I.

A martingale version of BMO can be found in [**8**], and [**1**] studies BMO on spaces of homogeneous type.

Later D. Sarason [**9**] introduced a subspace of BMO functions called VMO functions. In his paper [**9**], he gave several characterizations of this subspace, including that it is the closure in BMO of the set of uniformly continuous functions in BMO. Roughly speaking, a function in BMO is said to have *Vanishing Mean Oscillation*, if its mean oscillation is locally small, in a uniform sense. More precisely, a BMO function f is in VMO if $I(|f - I(f)|)$ tends to zero as $|I|$ tends to zero. We refer the reader to [**10**] and [**3**] for beautiful accounts of the spaces of BMO and VMO.

2. BMO Triples

We call a triple $(X, \mu, \mathcal{I})$ a *BMO triple* if X is a complete separable metric space with no isolated points, μ is a nonatomic regular Borel measure on X whose support is X (i.e., if $U \neq \varnothing$ is open in X, then $\mu(U) > 0$), and $\mathcal{I}$ is a collection of Borel subsets of X such that:

(1) $0 < \mu(I) < \infty$ for every $I \in \mathcal{I}$,
(2) for every nonempty open set $U \subset X$ there is an $I \in \mathcal{I}$ such that $I \subset U$,
(3) there is a countable subset $\{I_n\}_{n\geq 1}$ of $\mathcal{I}$ such that

$$\bigcup_{n\geq 1} I_n = X,$$

(4) for every I and J in $\mathcal{I}$, there are $I = I_1, I_2 \ldots, I_n = J$ in $\mathcal{I}$ such that, for each j, $1 \leq j < n$, either $I_j \subset I_{j+1}$ or $I_{j+1} \subset I_j$.

REMARK 2.1. If in (3) above we have $I_1 \subset I_2 \subset \cdots$ and if we have that, whenever $I, J \in \mathcal{I}$ and $\mu(I \cap J) > 0$, there is an $E \in \mathcal{I}$ with $E \subset I \cap J$, then we get statement (4) for free. These conditions hold in most of the classical examples.

Throughout this paper $(X, \mu, \mathcal{I})$ will denote a BMO triple. We define $L^1_{\mathcal{I},loc}(\mu)$ to be the collection of all measurable functions $f : X \to \mathbb{C}$ such that $\int_I |f|\, d\mu < \infty$ for every $I \in \mathcal{I}$. For $f \in L^1_{\mathcal{I},loc}(\mu)$ we define the *average* of f over I by

$$I(f) = \frac{1}{\mu(I)} \int_I f d\mu,$$

we define the *mean oscillation* of f on I by $I(|f - I(f)|)$, and we define

$$\|f\|^*_{BMO(\mathcal{I},\mu)} = \sup_{I\in\mathcal{I}} I(|f - I(f)|).$$

We define $BMO(\mathcal{I},\mu) = \left\{ f \in L^1_{\mathcal{I},loc}(\mu) : \|f\|^*_{BMO(\mathcal{I},\mu)} < \infty \right\}$. We also define the space $VMO(\mathcal{I},\mu)$ to be the set of all functions $f \in BMO(\mathcal{I},\mu)$ such that

$$\lim_{\mu(I)+diam(I)\longrightarrow 0} I\left(|f - I(f)|\right) = 0.$$

It is clear that $\|f\|^*_{BMO(\mathcal{I},\mu)} = 0$ if and only if the function f is constant *a.e.* (μ) on every $I \in \mathcal{I}$, and conditions (3) and (4) on $\mathcal{I}$ imply that $\|f\|^*_{BMO(\mathcal{I},\mu)} = 0$ if and only if the function f is constant *a.e.* (μ) on X.

DEFINITION 2.2. Suppose f and $\mathcal{I}$ are as above. If $X \in \mathcal{I}$, we define

$$\|f\|_{BMO(\mathcal{I},\mu)} = \|f\|^*_{BMO(\mathcal{I},\mu)} + \frac{1}{\mu(X)}\left|\int_X f d\mu\right|.$$

Otherwise, we define

$$\|f\|_{BMO(\mathcal{I},\mu)} = \|f\|^*_{BMO(\mathcal{I},\mu)},$$

and in this case, to make $\|f\|_{BMO(\mathcal{I},\mu)}$ a norm, we identify functions in $BMO(\mathcal{I},\mu)$ that differ by a constant, i.e., we mod out by the subspace of constant functions.

It is apparent that $f \in BMO(\mathcal{I},\mu)$ if and only if $\mathrm{Re}(f)$, $\mathrm{Im}(f) \in BMO(\mathcal{I},\mu)$. It is also simple but a useful fact that the space of real-valued $BMO(\mathcal{I},\mu)$ functions forms a lattice. In other words, if the real-valued functions $f, g \in BMO(\mathcal{I},\mu)$, then $|f|, |g| \in BMO(\mathcal{I},\mu)$, and therefore, so do $\max(f,g)$ and $\min(f,g)$.

The reader should note that this notion of BMO includes all classical BMO definitions [**7**], [**3**],[**10**].

LEMMA 2.3. *Suppose $\{f_n\}$ is a Cauchy sequence in $BMO(\mathcal{I},\mu)$ and $J \in \mathcal{I}$. Let $g_n = f_n - J(f_n)$. Then $I(g_n)$ is a Cauchy sequence for every $I \in \mathcal{I}$.*

PROOF. The proof when $J \subset I$ follows from the following:

$$|I(g_n) - I(g_m)| = |I(f_n - f_m) - J(f_n - f_m)| \leq J\left(|f_n - f_m - I(f_n - f_m)|\right)$$

$$\leq \frac{\mu(I)}{\mu(J)} I\left(|f_n - f_m - I(f_n - f_m)|\right) \leq \frac{\mu(I)}{\mu(J)} \|f_n - f_m\|_{BMO(\mathcal{I},\mu)}.$$

The proof when $I \subset J$ is the same. For the general case choose $I_1 = I, \cdots, I_M = J$ in $\mathcal{I}$ as in condition (4) in the definition of $\mathcal{I}$, and note that

$$|I(g_n - g_m)| = |I(g_n - g_m) - J(g_n - g_m)| \leq \sum_{k=1}^{M-1} |I_k(g_n - g_m) - I_{k+1}(g_n - g_m)|$$

$$\leq \|f_n - f_m\|_{BMO(\mathcal{I},\mu)} \sum_{k=1}^{M-1}\left[\frac{\mu(I_k)}{\mu(I_{k+1})} + \frac{\mu(I_{k+1})}{\mu(I_k)}\right].$$

□

PROPOSITION 1. $\left(BMO(\mathcal{I},\mu), \|\cdot\|_{BMO(\mathcal{I},\mu)}\right)$ *is a Banach space.*

PROOF. We only need to show that $BMO(\mathcal{I},\mu)$ is complete. Suppose $\{f_n\}$ is a Cauchy sequence in $BMO(\mathcal{I},\mu)$. First suppose that $X \notin \mathcal{I}$. Fix $I_0 \in \mathcal{I}$ and let $g_n = f_n - I_0(f_n)$. For every $I \in \mathcal{I}$ and $f \in BMO(\mathcal{I},\mu)$ we have:

$$\begin{aligned}\|f\|_{1,I} &= \int_I |f|\,d\mu \leq \int_I |f - I(f)|\,d\mu + |I(f)|\,\mu(I) \\ &\leq \mu(I)\,\|f\|_{BMO(\mathcal{I},\mu)} + |I(f)|\,\mu(I).\end{aligned}$$

Since $I(|g_n - g_m - I(g_n - g_m)|) = I(|f_n - f_m - I(f_n - f_m)|)$ we have $\|g_n - g_m\|_{BMO(\mathcal{I},\mu)} = \|f_n - f_m\|_{BMO(\mathcal{I},\mu)}$. From Lemma 2.3 we know that $\{I\,(g_n)\}$ is a Cauchy sequence. Thus the above inequality with $f = g_n - g_m$ implies that $\{g_n\}$ is Cauchy in $L^1\,(I)$ for every I and so is convergent in L^1-norm to a function $g \in L^1\,(I)$. We have:

$$\begin{aligned}& I(|g_n - g - I(g_n - g)|) \\ \leq\ & I(|g_n - g_m - I(g_n - g_m)|) + I(|g_m - g - I(g_m - g)|) \\ \leq\ & \|g_n - g_m\|_{BMO(\mathcal{I},\mu)} + 2I(|g_m - g|) \\ =\ & \|g_n - g_m\|_{BMO(\mathcal{I},\mu)} + \frac{2}{\mu(I)}\,\|g_m - g\|_{1,I}\,.\end{aligned}$$

Suppose $\varepsilon > 0$ is given. There exists N such that $m, n \geq N$ implies that (from above):

$$I(|g_n - g - I(g_n - g)|) < \frac{\varepsilon}{2} + \frac{2}{\mu(I)}\,\|g_m - g\|_{1,I}\,.$$

By letting $m \longrightarrow \infty$ it follows that $I(|\ g_n - g - I(g_n - g)\ |) \leq \frac{\varepsilon}{2}$ for every I. Therefore g_n converges to g in $BMO(\mathcal{I},\mu)$ norm. Since $X \notin \mathcal{I}$, then $f_n = g_n$ in $BMO(\mathcal{I},\mu)$ and so f_n is convergent to g.

If $X \in \mathcal{I}$, we have

$$\begin{aligned}X\,(|f_n - f_m|) &\leq X\,(|f_n - f_m - X\,(f_n - f_m)|) + |X\,(f_n - f_m)| \\ &\leq \|f_n - f_m\|_{BMO(\mathcal{I},\mu)}\end{aligned}$$

thus f_n converges in L^1-norm to a function $f \in L^1\,(X)$. The proof of convergence of g_n can be applied to show that f_n converges to f in $BMO(\mathcal{I},\mu)$ norm. □

In the proof of the next corollary we have used the ideas of the proof of the preceding proposition.

COROLLARY 1. *Let $(X,\mu,\mathcal{I})$ and $(X,\mu,\mathcal{J})$ be two BMO triples and $BMO(\mathcal{I},\mu) \subset BMO(\mathcal{J},\mu)$. Then there exists $M > 0$ such that $\|f\|_{BMO(\mathcal{J},\mu)} \leq M\,\|f\|_{BMO(\mathcal{I},\mu)}$, $\forall f \in BMO(\mathcal{I},\mu)$. In particular, $\|\cdot\|_{BMO(\mathcal{I},\mu)}$ is equivalent to $\|\cdot\|_{BMO(\mathcal{J},\mu)}$ if and only if $BMO(\mathcal{I},\mu) = BMO(\mathcal{J},\mu)$.*

PROOF. Let $\varphi : BMO(\mathcal{I},\mu) \longrightarrow BMO(\mathcal{J},\mu)$ be the identity map. By using the Closed-Graph Theorem we will show that φ is a linear bounded map. Suppose that $f_n \in BMO(\mathcal{I},\mu)$, $f_n \longrightarrow f$ in $BMO(\mathcal{I},\mu)$, and that $f_n \longrightarrow g$ in $BMO(\mathcal{J},\mu)$. We will show that $f = g$ in $BMO(\mathcal{I},\mu)$. It is clear that if $\mu\,(X) < \infty$, then $BMO(\mathcal{I},\mu) = BMO(\mathcal{I} \cup \{X\},\mu)$ and $BMO(\mathcal{J},\mu) = BMO(\mathcal{J} \cup \{X\},\mu)$. Thus without loss of generality we can assume that $X \in \mathcal{I} \cap \mathcal{J}$ whenever $\mu\,(X) < \infty$. The rest of the proof divides into two cases. First suppose that $\mu\,(X) < \infty$. Then, similar to the proof of the previous proposition, we have $X\,(|f_n - f|) \leq \|f_n - f\|_{BMO(\mathcal{I},\mu)}\,.$

Thus $f_n \longrightarrow f$ in $L^1(X)$. In the same way, $f_n \longrightarrow g$ in $L^1(X)$. Therefore $f = g$ almost everywhere. Next suppose that $\mu(X) = \infty$. Choose $I^{'} \in \mathcal{I}$ and $J^{'} \in \mathcal{J}$ such that $\mu\left(I^{'} \cap J^{'}\right) > 0$ and let $I_0 = I^{'} \cap J^{'}$. Without loss of generality we can assume that $I_0 \in \mathcal{I} \cap \mathcal{J}$. By the proof of the previous proposition, it follows that $f_n - I_0(f_n) \longrightarrow f - I_0(f)$ in $L^1(I)$ for every $I \in \mathcal{I}$. A similar proof shows that $f_n - I_0(f_n) \longrightarrow g - I_0(g)$ in $L^1(J)$ for every $J \in \mathcal{J}$. Thus $f - I_0(f) = g - I_0(g)$ on $I \cap J$, almost everywhere, $\forall I \in \mathcal{I}$ and $\forall J \in \mathcal{J}$. Since $X = \cup_{n \geq 1} I_n = \cup_{n \geq 1} J_n$ for some $I_n \in \mathcal{I}$ and $J_n \in \mathcal{J}$, it follows that $f - I_0(f) = g - I_0(g)$ almost everywhere on X. Therefore $f = g$ in $BMO(\mathcal{I}, \mu)$. □

Let $C_u(X)$ denote the set of uniformly continuous functions on X. If $I \in \mathcal{I}$, define the measure μ_I as the restriction of μ to the σ-algebra of Borel subsets of I.

LEMMA 2.4. *If $(X, \mu, \mathcal{I})$ is a BMO triple, then:*

1. *$C_u(X) \cap BMO(\mathcal{I}, \mu) \subset VMO(\mathcal{I}, \mu)$ and $VMO(\mathcal{I}, \mu)$ is a closed linear subspace of $BMO(\mathcal{I}, \mu)$.*
2. *There is a countable collection of continuous linear functionals on $BMO(\mathcal{I}, \mu)$ that separates the points of $BMO(\mathcal{I}, \mu)$.*
3. *For every $f \in L^\infty(\mu)$,*
$$\|f\|_{BMO(\mathcal{I},\mu)} \leq 3 \|f\|_\infty .$$
In particular, the inclusion map from $L^\infty(\mu)$ to $BMO(\mathcal{I}, \mu)$ is continuous.

PROOF. (1) The inclusion $C_u(X) \cap BMO(\mathcal{I}, \mu) \subset VMO(\mathcal{I}, \mu)$ is easily proved. For each $I \in \mathcal{I}$, we define $T_I : BMO(\mathcal{I}, \mu) \to L^1(\mu_I)$ by
$$T_I(f) = \frac{1}{\mu(I)} (f - I(f))|_I .$$
Then $\|T_I\| \leq 1$ and $f \in VMO(\mathcal{I}, \mu)$ if and only if
$$\lim_{\mu(I)+diam(I) \to 0} \|T_I(f)\| = 0.$$
It easily follows that $VMO(\mathcal{I}, \mu)$ is a closed linear subspace of $BMO(\mathcal{I}, \mu)$.

(2) For every I_n in the definition of BMO triple there exist continuous linear functionals $\{\phi_{n,k}\}_{k \geq 1}$ on $L^1\left(\mu_{I_n}\right)$ that separate the points of $L^1\left(\mu_{I_n}\right)$. Define $\psi_{n,k} : BMO(\mathcal{I}, \mu) \to \mathbb{C}$ by
$$\psi_{n,k}(f) = \phi_{n,k}\left(\frac{1}{\mu(I_n)} (f - I_n(f))|_{I_n}\right).$$
Note that
$$|\psi_{n,k}(f)| \leq \|\phi_{n,k}\| \frac{1}{\mu(I_n)} \|(f - I_n(f))|_{I_n}\|_1 \leq \|\phi_{n,k}\| \|f\|_{BMO(\mathcal{I},\mu)} .$$
It now follows that $\{\psi_{n,k} : n, k \in \mathbb{N}\}$ separates the points of $BMO(\mathcal{I}, \mu)$.

(3) This is obvious.

□

If X is the unit circle, μ is the normalized arc length, and $\mathcal{I}$ is the set of all arcs in X, then we obtain the classical BMO and VMO spaces defined on the unit circle. The following proposition shows that our general versions can be quite different.

PROPOSITION 2. *If* $\mathcal{I} = \{I \subset X : 0 < \mu(I) < \infty,\ I \text{ is a Borel set}\}$, *then:*

(1) $BMO(\mathcal{I}, \mu) = L^\infty(\mu)$,

(2) *if* $L = \min\{\|f - \lambda\|_\infty : \lambda \in \mathbb{C}\}$, *then* $L \leq \|f\|_{BMO(I,\mu)} \leq 3\|f\|_\infty$ *for every* $f \in L^\infty(\mu)$,

(3) $VMO(\mathcal{I}, \mu) = C_u(X) \cap L^\infty(\mu)$.

PROOF. 1) Suppose $f \notin L^\infty(\mu)$. Then, for each positive integer n there are complex numbers a_1, a_2 with $|a_1 - a_2| \geq 2n + 2$ such that

$$\mu(\{x \in X : |f(x) - a_j| < 1\}) > 0 \text{ for } j = 1, 2.$$

Since μ is σ-finite and nonatomic, there are subsets $E_j \subset \{x \in X : |f(x) - a_j| < 1\}$ for $j = 1, 2$ such that $0 < \mu(E_1) = \mu(E_2) < \infty$. If we let $I = E_1 \cup E_2$, then

$$\left|\frac{a_1 + a_2}{2} - I(f)\right| = \left|\frac{\frac{1}{\mu(E_1)}\int_{E_1}(a_1 - f)\,d\mu + \frac{1}{\mu(E_2)}\int_{E_2}(a_2 - f)\,d\mu}{2}\right| < 1,$$

$$\|f\|_{BMO(\mathcal{I},\mu)} \geq \frac{1}{\mu(I)}\int_I |f - I(f)|\,d\mu \geq n.$$

2) We assume that f is not a constant function. Suppose also that f is a real-valued function and let M and m be the *essential supremum* and the *essential infimum* of the function, i.e., $m \leq |f(x)| \leq M$ a.e. μ and $\mu\{x : M - \varepsilon < f(x) \leq M\} > 0$ and $\mu\{x : m \leq f(x) < m + \varepsilon\} > 0$ for every $\varepsilon > 0$. For every positive integer n, there exist $\varepsilon_n > 0$, $\delta_n > 0$, and Borel subsets $I_{1,n}$ and $I_{2,n}$ with the following properties: $\mu(I_{1,n}) \approx \mu(I_{2,n}) < \infty$, $M - \varepsilon_n \leq f(x) \leq M$ for every $x \in I_{1,n}$, and $m \leq f(y) \leq m + \delta_n$ for every $y \in I_{1,n}$. We can choose ε_n and δ_n so that they both converge to zero as n goes to infinity. Let $I_n = I_{1,n} \cup I_{2,n}$. It follows that $\frac{1}{2}(m + M - \varepsilon_n) \leq I_n(f) \leq \frac{1}{2}(M + m + \delta_n)$ for every $n \geq 1$. Therefore:

$$\begin{aligned}
I_n(|f - I_n(f)|) &= \frac{1}{\mu(I_n)}\left[\int_{I_{1,n}} |f - I_n(f)|\,d\mu + \int_{I_{2,n}} |f - I_n(f)|\,d\mu\right] \\
&= \frac{1}{\mu(I_n)}\left[\int_{I_{1,n}} (f - I_n(f))d\mu + \int_{I_{2,n}} (I_n(f) - f)d\mu\right] \\
&\approx \frac{1}{\mu(I_n)}\left[\int_{I_{1,n}} f d\mu - \int_{I_{2,n}} f d\mu\right].
\end{aligned}$$

The proof for this case will be completed by noticing that $L = \frac{1}{2}(M - m)$ and the following inequalities:

$$\frac{1}{2}(M - m - \delta_n - \varepsilon_n) \leq \frac{1}{\mu(I_n)}\left[\int_{I_{1,n}} f d\mu - \int_{I_{2,n}} f d\mu\right] \leq \frac{1}{2}(M - m).$$

The proof of the general case that f is a complex valued function will be apparent if one uses the previous case and the facts that $\|f\|_\infty = \sup_{0 \leq \theta \leq 2\pi} \left\|\mathrm{Re}(e^{i\theta} f)\right\|_\infty$ and that $\|\mathrm{Re}(g)\|_{BMO(\mathcal{I},\mu)} \leq \|g\|_{BMO(\mathcal{I},\mu)}$ for every $g \in BMO(\mathcal{I}, \mu)$.

3) First we will show that $VMO(\mathcal{I}, \mu) \subset C(X)$. Suppose $f \in BMO(\mathcal{I}, \mu) \backslash C(X)$ is real valued. By a theorem of D. Hadwin [**5**], there exists a point $a \in X$ such that f cannot be continuous at $x = a$ by redefining the function on any set of measure zero. For every $n \geq 1$ define the monotone sequences M_n and m_n to be the essential

supremum and essential infimum of the function f on the open disk centered at a and radius $\frac{1}{n}$, $\mathcal{B}\left(a; \frac{1}{n}\right)$, respectively. The sequence $M_n - m_n$ does not converge to zero, otherwise f would be continuous at $x = a$ by redefining it at $x = a$ to be $\lim_{n \longrightarrow \infty} M_n$. Therefore there exists $\varepsilon > 0$ such that, without loss of generality, $M_n - m_n \geq \varepsilon$ for every $n \geq 1$. Choose Borel subsets $I_{n,1}$ and $I_{n,2}$ of $\mathcal{B}\left(a; \frac{1}{n}\right)$, of equal measure such that $f(t_2) - f(t_1) > \frac{\varepsilon}{2}$ for every $t_1 \in I_{n,1}$ and $t_2 \in I_{n,2}$. Thus $I_{n,2}(f) - I_{n,1}(f) \geq \frac{\varepsilon}{2}$. By letting $I_n = I_{n,1} \cup I_{n,2}$ we will have:

$$\begin{aligned} I_n\left(|f - I_n(f)|\right) &\geq \frac{1}{2} I_{n,2}\left(|f - I_n(f)|\right) \geq \frac{1}{2}\left|I_{n,2}\left(f - I_n(f)\right)\right| \\ &= \frac{1}{2}\left|I_{n,2}(f) - I_n(f)\right| = \frac{1}{2}\left|I_{n,2}(f) - I_{n,1}(f)\right| \\ &\geq \frac{\varepsilon}{4}. \end{aligned}$$

Therefore $f \notin VMO(\mathcal{I}, \mu)$.

We have shown that

$$C_u(X) \cap BMO(\mathcal{I}, \mu) \subset VMO(\mathcal{I}, \mu) \subset C(X).$$

To finish the proof suppose that $g \in VMO(\mathcal{I}, \mu) \backslash C_u(X)$. Thus there exists a continuous function f on X in $BMO(\mathcal{I}, \mu)$ such that $f = g$ almost everywhere. There exist $x_n, y_n \in X$ and $\varepsilon > 0$ such that $d\left(x_n, y_n\right) \longrightarrow 0$ but $\left|f\left(x_n\right) - f\left(y_n\right)\right| \geq \varepsilon$ for every $n \geq 1$. The continuity of f at x_n and y_n implies that there are Borel sets $x_n \in I_{n,1}$ and $y_n \in I_{n,2}$ in small neighborhood of x_n and y_n such that $\mu\left(I_{n,1}\right) = \mu\left(I_{n,2}\right) \longrightarrow 0$. If we let $I_n = I_{n,1} \cup I_{n,2}$ and do a similar calculations as above, then we arrive to the contradiction that $f \notin VMO(\mathcal{I}, \mu)$. □

3. Characterizations of VMO

Here is a natural question that arises from the classical case and the preceding proposition:

Is $VMO(\mathcal{I}, \mu)$ the $\|\cdot\|_{BMO(\mathcal{I}, \mu)}$-closure of $C_u(X) \cap BMO(\mathcal{I}, \mu)$?

We will give an affirmative answer to the above question in some special cases. The proof of the next lemma is a small modification of [**9**, Lemma 2]. We present the proof here for the sake of completeness.

LEMMA 3.1. *Suppose there exists $M \geq 1$ for which for every $n \geq 1$ there exists $\mathcal{J}_n \subset \mathcal{I}$ that partitions X and satisfies in the following conditions:*

(1) $\mu(J) + diam(J) \leq \frac{1}{n}$ *for every* $J \in \mathcal{J}_n$,

(2) $\forall I_1, I_2 \in \mathcal{J}_n$ *if* $I_1 \cup I_2 \subset \mathcal{B}\left(x; \frac{3}{n}\right)$ *for some* $x \in X$, *then* $\exists I \in \mathcal{I}$ *such that* $I_1 \cup I_2 \subset I$ *and*

$$\mu(I) \leq M\mu(I_k) \text{ and } diam(I) \leq M diam(I_k) \text{ for } k = 1, 2$$

(3) *If* $I \in \mathcal{I}$ *and* $\frac{1}{n} \leqslant \mu(I) + diam(I)$, *then there are finitely many* $I_1, \cdots, I_L \in \mathcal{J}_n$ *such that* $I \subset I_1 \cup \cdots \cup I_L$ *a.e. and* $\mu(I_1 \cup \cdots \cup I_L) \leq M\mu(I)$.

Then f_n defined by $f_n(x) = \sum_{J \in \mathcal{J}_n} J(f)\chi_J(x)$ *converges to f in the BMO-norm for every function* $f \in VMO(\mathcal{I}, \mu)$.

PROOF. For given $\varepsilon > 0$ there exists $\delta > 0$ such that $I(|f - I(f)|) < \varepsilon$ whenever $I \in \mathcal{I}$ and $\mu(I) + diam(I) < \delta$. Suppose $n \in \mathbb{N}$ and $\frac{1}{n} < \frac{\delta}{M}$. We claim that

$$\forall x, y \in X \text{ if } d(x, y) < \frac{1}{n}, \text{ then } |f_n(x) - f_n(y)| < 2M\varepsilon.$$

To see this, assume that $x \in I_1$ and $y \in I_2$ where $I_1, I_2 \in \mathcal{J}_n$. Since $I_1 \cup I_2 \subset \mathcal{B}\left(x; \frac{3}{n}\right)$, there exists $I \in \mathcal{I}$ that satisfies in the condition (2) of the lemma. Thus we have

$$\begin{aligned} |f_n(x) - f_n(y)| &\leq |f_n(x) - I(f)| + |I(f) - f_n(y)| \\ &\leq I_1(|f - I(f)|) + I_2(|f - I(f)|) \\ &\leq \frac{\mu(I)}{\mu(I_1)} I(|f - I(f)|) + \frac{\mu(I)}{\mu(I_2)} I(|f - I(f)|) \\ &< 2M\varepsilon. \end{aligned}$$

To estimate $\|f - f_n\|_{BMO(\mathcal{I},\mu)}$, suppose $I \in \mathcal{I}$. If $\mu(I) + diam(I) < \frac{1}{n}$, then

$$\begin{aligned} I(|f - f_n - I(f - f_n)|) &\leq I(|f - I(f)|) + \frac{1}{\mu(I)^2} \int_I \int_I |f_n(x) - f_n(y)| \, d\mu d\mu \\ &< \varepsilon + 2M\varepsilon \leq 3M\varepsilon. \end{aligned}$$

If $\mu(I) + diam(I) \geq \frac{1}{n}$ there are $I_1, \cdots, I_L \in \mathcal{J}_n$ that satisfy in condition (3) of the lemma. Then

$$\begin{aligned} I(|f - f_n - I(f - f_n)|) &\leq 2I(|f - f_n|) \leq \frac{2}{\mu(I)} \sum_{i=1}^{L} \int_{I_i} |f - f_n| \\ &= \frac{2}{\mu(I)} \sum_{i=1}^{L} \int_{I_i} |f - I_i(f)| < \frac{2}{\mu(I)} \sum_{i=1}^{L} \mu(I_i) \varepsilon \\ &\leq 2M\varepsilon. \end{aligned}$$

Therefore for every $I \in \mathcal{I}$ we have $I(|f - f_n - I(f - f_n)|) \leq 3M\varepsilon$ which implies that $\|f - f_n\|_{BMO(\mathcal{I},\mu)} \leq 9M\varepsilon$. □

EXAMPLE 3.2. Suppose $A \neq \emptyset$ is any open subset of $\mathbb{R}^m$ which is bounded and convex. Let $\mathcal{I} = \{v + \alpha A : v \in \mathbb{R}^m, \alpha \in \mathbb{R}^+\}$ and for each $n \in \mathbb{Z}^+$ let

$$\mathcal{J}_n = \left\{ v + \frac{1}{n(\mu(A) + diam(A))} A : v \in \mathbb{R}^m \right\}.$$

Then we can show that $\mathcal{I}$ and $\mathcal{J}_n$ satisfy the conditions of the previous lemma. To see this, let $I \in \mathcal{I}$. To simplify the calculations we assume that $\mu(A) + diam(A) = 1$. Suppose A' and I' are the largest cubes contained in A and I, respectively. If we need N many cubes of the form $v + A'$ to cover A, then N many cubes of the form $v + I'$ would cover I and N many cubes of the form $J'_{n,v} = v + \frac{1}{n} A' \subset J_{n,v} = v + \frac{1}{n} A \in \mathcal{J}_n$ would cover any element of $\mathcal{J}_n$. Let $\beta = \dfrac{\mu(I)}{\mu(A)} = \dfrac{\mu(I')}{\mu(A')}$. Then

$$\frac{\mu(I)}{\mu(J_{n,v})} = \frac{\mu(I')}{\mu(J'_{n,v})} = \frac{\beta \mu(A')}{\left(\frac{1}{n}\right)^m \mu(A')} = \beta n^m.$$

Note that $\mu(I) + diam(I) \geqslant \frac{1}{n} \geqslant \mu(J_{n,v}) + diam(J_{n,v})$ implies that $\mu(I) \geqslant \mu(J_{n,v})$, and so $\beta n^m = \dfrac{\mu(I)}{\mu(J_{n,v})} \geqslant 1$. Each $v+I'$ can be covered by $2^m(\lfloor \beta n^m \rfloor + 1)$ $J'_{n,v}$'s, so I can be covered by at most $2^m N(\lfloor \beta n^m \rfloor + 1)$ $J_{n,v}$'s. Therefore:

$$\begin{aligned}
\frac{\mu(J_{n,v}\text{'s})}{\mu(I)} &\leqslant \frac{2^m N(\lfloor \beta n^m \rfloor + 1)\mu\left(\frac{1}{n}A\right)}{\beta\mu(A)} \\
&\leqslant 2^m N\left(1 + \frac{2}{\beta n^m}\right) \\
&\leqslant 2^m 3N.
\end{aligned}$$

The next lemma states a condition under which we can approximate a function by a continuous function.

LEMMA 3.3. *Suppose that Y is a Hausdorff paracompact topological space, $\varepsilon > 0$, $f : Y \to \mathbb{C}$, and $\mathcal{U}$ is an open cover of Y such that, for every $U \in \mathcal{U}$ and every $x, y \in U$ we have $|f(x) - f(y)| < \varepsilon$. Then there is a continuous function $g : Y \to \mathbb{C}$ such that, for every $x \in Y$,*

$$|f(x) - g(x)| < \varepsilon.$$

PROOF. Let $\{g_\lambda : \lambda \in \Lambda\}$ be a partition of unity subordinate to $\mathcal{U}$, and for each $\lambda \in \Lambda$ choose $x_\lambda \in supp(g_\lambda)$. We define $g : Y \to \mathbb{C}$ by

$$g(x) = \sum_{\lambda\in\Lambda} f(x_\lambda) g_\lambda(x).$$

Suppose $x \in Y$ and let $D = \{\lambda \in \Lambda : g_\lambda(x) > 0\}$. From the definition of a partition of unity there is a $U \in \mathcal{U}$, an open set V such that $x \in V \subset U$ and $\sum_{\lambda\in D} g_\lambda(y) = 1$ for every $y \in V$ and $\sum_{\lambda\in D} g_\lambda(y) = 0$ for every $y \in Y \backslash U$. Hence, for every $\lambda \in D$, $x_\lambda \in U$. Thus, for every $\lambda \in D$ we have $|f(x) - f(x_\lambda)| < \varepsilon$. It follows that g is continuous on V (and by the generality of x, on Y) and that

$$|f(x) - g(x)| = \left|\sum_{\lambda\in D} (f(x) - f(x_\lambda)) g_{\lambda(x)}\right| < \varepsilon.$$

□

COROLLARY 2. *Suppose (Y, d) is a compact metric space, $f : Y \to \mathbb{C}$, $\varepsilon > 0$, and for every $x \in Y$ there is a $\delta_x > 0$ such that whenever $y \in Y$ and $d(x, y) < \delta_x$ we have $|f(x) - f(y)| < \varepsilon$. Then there is a uniformly continuous function $g : Y \to \mathbb{C}$ such that*

$$|f(x) - g(x)| < \varepsilon$$

for every $x \in Y$.

The preceding corollary uses compactness to guarantee that the continuous function g is actually uniformly continuous. If $Y \subset \mathbb{R}^n$, we can obtain uniform continuity with a weaker hypothesis.

PROPOSITION 3. *Suppose E is a subset of $\mathbb{R}^n$, $\varepsilon > 0, \delta > 0$ and $\varphi : E \to \mathbb{C}$ is a function such that, for every $x, y \in E$,*

$$\|x - y\|_\infty < \delta \Rightarrow |\varphi(x) - \varphi(y)| < \varepsilon.$$

Then there is a uniformly continuous function $F : E \to \mathbb{C}$ such that

$$|\varphi(x) - F(x)| < \varepsilon$$

for every $x \in E$.

PROOF. Let $\omega = \delta/4$. For each integer $k \in \mathbb{Z}$ we define

$$f_k(t) = \begin{cases} 0 & \text{if } t < (2k-1)\omega \\ \frac{1}{\omega}(t - (2k-1)\omega) & \text{if } (2k-1)\omega \le t < 2k\omega \\ 1 & \text{if } 2k\omega \le t \le (2k+1)\omega \\ -\frac{1}{\omega}(t - (2k+1)\omega) & \text{if } (2k+1)\omega < t \le (2k+2)\omega \\ 0 & \text{if } (2k+2)\omega < t \end{cases}$$

It is easily seen that

(1) $|f_k(s) - f_k(t)| \le \frac{1}{\omega}|s - t|$ for every $k \in \mathbb{Z}$ and all $s, t \in \mathbb{R}$, and
(2) $\sum_{k \in \mathbb{Z}} f_k(t) = 1$ for every $t \in \mathbb{R}$.

If $\kappa = (k_1, \ldots, k_n) \in \mathbb{Z}^n$, we define $f_\kappa : \mathbb{R}^n \to \mathbb{C}$ by

$$f_\kappa(t_1, \ldots, t_n) = \prod_{j=1}^n f_{k_j}(t_j).$$

It is easy to show,

3. for all $s, t \in \mathbb{R}^n$ with $\|s - t\|_\infty < 1$, that

$$\begin{aligned} |f_\kappa(s) - f_\kappa(t)| &= \left| \prod_{j=1}^n \left[\left(f_{k_j}(s_j) - f_{k_j}(t_j) \right) + f_{k_j}(t_j) \right] - \prod_{j=1}^n f_{k_j}(t_j) \right| \\ &\le \left(\frac{2}{\omega} \right)^n \|s - t\|_\infty, \end{aligned}$$

4. if $x \in \mathbb{R}^n$,

$$\text{ball}_{\|\,\|_\infty}(x, \omega) \subset \bigcup_{x \in supp(f_\kappa)} supp(f_\kappa) \subset \text{ball}_{\|\,\|_\infty}(x, 3\omega),$$

5. $\sum_{\kappa \in \mathbb{Z}^n} f_\kappa(t) = 1$ for every $t \in \mathbb{R}^n$.

Define $\Lambda = \{\kappa \in \mathbb{Z}^n : \exists x_\kappa \in E \text{ with } dist(x_\kappa, supp(f_\kappa)) < 3\omega\}$, and we define $F : \mathbb{R}^n \to \mathbb{C}$ by

$$F(t) = \sum_{\kappa \in \Lambda} \varphi(x_\kappa) f_\kappa(t).$$

Suppose $x \in E$. It follows from (4) above, for each $\kappa \in \mathbb{Z}^n$ with $x \in supp(f_\kappa)$, that $\|x - x_\kappa\|_\infty < 3\omega < \delta$, which implies by hypothesis that $|\varphi(x_\kappa) - \varphi(x)| < \varepsilon$. It now follows from (5) that

$$|F(x) - \varphi(x)| = \left| \sum_{\kappa \in \Lambda, x \in supp(f_\kappa)} [\varphi(x_\kappa) - \varphi(x)] f_{\kappa(x)} \right| < \varepsilon.$$

To show that F is uniformly continuous on E, suppose $x, y \in E$ and $\|x-y\|_\infty < \min(1,\omega)$. Let D be the set of all κ such that x or y is in $supp(f_\kappa)$. It follows from (4) that, whenever $\kappa \in D$,

$$\|x - x_\kappa\|_\infty < 4\omega < \delta,$$

which implies

$$|\varphi(x) - \varphi(x_\kappa)| < \varepsilon.$$

Hence, by (5)

$$|F(x) - F(y)| = |\sum_{\kappa \in D} [\varphi(x_\kappa) - \varphi(x)] [f_\kappa(x) - f_\kappa(y)],$$

and, by (3) and the fact that the cardinality of D is at most $2(3^n)$,

$$|F(x) - F(y)| \leq 2(3^n)\varepsilon \left(\frac{2}{\omega}\right)^n \|s-t\|_\infty .$$

Hence F is uniformly continuous on E. □

We can now prove a generalization of Sarason's theorem [**9**].

THEOREM 3.4. *If X is a compact space or X is any subset of $\mathbb{R}^n$ and $\mathcal{I}$ satisfies the conditions of Lemma 3.1, then* $VMO(\mathcal{I},\mu) = \overline{C_u(X) \cap BMO(\mathcal{I},\mu)}^{\|\cdot\|_{BMO(\mathcal{I},\mu)}}$.

PROOF. Suppose $f \in VMO(\mathcal{I},\mu)$, X is compact, and $\varepsilon > 0$. Then by Lemma 3.1 there exists f_n as in Lemma 3.1 such that $\|f - f_n\|_{BMO(\mathcal{I},\mu)} < \frac{\varepsilon}{2}$. It was shown in the proof of Lemma 3.1 that $|f_n(x) - f_n(y)| < \frac{\varepsilon}{6}$ whenever $d(x,y)$ is small enough. Lemma 3.3 can be applied to find a continuous function h_n such that $\|f_n - h_n\|_\infty \leq \frac{\varepsilon}{6}$ and so $\|f_n - h_n\|_{BMO(\mathcal{I},\mu)} \leq \frac{\varepsilon}{2}$. Therefore $\|f - h_n\|_{BMO(\mathcal{I},\mu)} < \varepsilon$. If $X \subset \mathbb{R}^n$, then Lemma 3.1 and the previous proposition will imply that $VMO(\mathcal{I},\mu) = \overline{C_u(X) \cap BMO(\mathcal{I},\mu)}^{\|\cdot\|_{BMO(\mathcal{I},\mu)}}$. □

REMARK 3.5.

(1) If A and $\mathcal{I}$ are as in Example 3.2, then by the previous theorem $VMO(\mathcal{I},\mu) = \overline{C_u(X) \cap BMO(\mathcal{I},\mu)}^{\|\cdot\|_{BMO(\mathcal{I},\mu)}}$. As a special case that A is a ball or a cube in $\mathbb{R}^n$, then we get the Sarason's theorem [**9**].
(2) If X is a circle with $\mathcal{I}$ the set of open arcs or X is an interval with $\mathcal{I}$ the open subintervals, and if μ is a finite continuous (i.e., $\mu\{(x)\} = 0$ for every x) measure whose support is X, then the hypotheses of Lemma 3.1, so $VMO(\mathcal{I},\mu) = \overline{C_u(X) \cap BMO(\mathcal{I},\mu)}^{\|\cdot\|_{BMO(\mathcal{I},\mu)}}$.

The next theorem gives another characterization of $VMO(\mathcal{I},\mu)$. Following Sarason's notation [**9**], for a positive measurable function f and $a > 0$ we let:

$$N_a(f) = \sup_{\mu(I)+diam(I)\leq a} I(f)\, I\left(f^{-1}\right) \text{ and } N_0(f) = \lim_{a\longrightarrow 0} N_a(f).$$

The *Schwarz*'s inequality implies that $N_a(f) \geq 1$ for every $a > 0$. For the proof of the next theorem we refer the reader to [**9**].

THEOREM 3.6. *Let $f \in BMO(\mathcal{I},\mu)$ be a real valued function. Then $N_0\left(e^f\right) = 1$ if and only if $f \in VMO(\mathcal{I},\mu)$.*

It is clear that if $\mathcal{I} \subset \mathcal{J}$, then $BMO(\mathcal{J}, \mu) \subset BMO(\mathcal{I}, \mu)$ and $VMO(\mathcal{J}, \mu) \subset VMO(\mathcal{I}, \mu)$. The next result is a generalization of this.

PROPOSITION 4. *Let $(X, \mu, \mathcal{I})$ and $(X, \mu, \mathcal{J})$ be two BMO triples and $M \geq 1$. If for every $J \in \mathcal{J}$ there exists an $I \in \mathcal{I}$ such that $J \subset I$ and $\mu(I) \leq M\mu(J)$, then*

(1) $\|f\|_{BMO(\mathcal{J},\mu)} \leq 2M \|f\|_{BMO(\mathcal{I},\mu)}$,
(2) *if also $diam(I) \leq Mdiam(J)$, then $VMO(\mathcal{I}, \mu) \subset VMO(\mathcal{J}, \mu)$.*

PROOF. Suppose that $f \in BMO(\mathcal{I}, \mu)$ and $J \in \mathcal{J}$. Choose $I \in \mathcal{I}$ that contains J and $\mu(I) \leq M\mu(J)$. The proof will be apparent by the fact that:

$$\begin{aligned} J(\mid f - J(f) \mid) &\leq J(\mid f - I(f) \mid + \mid I(f) - J(f) \mid) \leq 2J(\mid f - I(f) \mid) \\ &\leq 2MI(\mid f - I(f) \mid). \end{aligned}$$

□

EXAMPLE 3.7. Let $X = \mathbb{R}^n$ and let μ be Lebesgue measure. If we let $\mathcal{I}$ be the collection of all disks in $\mathbb{R}^n$ and $\mathcal{J}$ be the collection of all cubes in $\mathbb{R}^n$, then by Proposition 4 and Corollary 1 $BMO(\mathcal{I}, \mu) = BMO(\mathcal{J}, \mu)$, $VMO(\mathcal{I}, \mu) = VMO(\mathcal{J}, \mu)$. More generally, if A and $\mathcal{K}$ are as in Example 3.2, then $BMO(\mathcal{I}, \mu) = BMO(\mathcal{K}, \mu)$ and $VMO(\mathcal{I}, \mu) = VMO(\mathcal{K}, \mu)$.

4. John-Nirenberg Theorem

In this part we will prove a version of John-Nirenberg theorem for any function $f \in BMO(\mathcal{I}, \mu)$ whenever $\mathcal{I}$ has certain properties. We let *essdiam*(J) denote the *essential diameter* of J (i.e., the diameter modulo sets of measure 0). **Note:** when we talk of a *partition*, we always mean *modulo sets of measure zero.*

A key idea in this section is the existence of certain partitions of sets. Suppose μ is a Borel measure on a metric space Y. A sequence $\{\mathcal{A}_n\}$ of measurable partitions of Y is called a *null sequence of partitions* if

(1) $\mathcal{A}_{n+1}$ is a refinement of $\mathcal{A}_n$ for every $n \geq 1$.
(2) If $\{J_n\}$ is a sequence and $J_n \in \mathcal{A}_n$ for every $n \geq 1$, then

$$\mu(J_n) + essdiam(J_n) \longrightarrow 0.$$

LEMMA 4.1. *Suppose μ is a Borel measure on a metric space Y, and suppose $\{\mathcal{A}_n\}$ is a null sequence of partitions of Y Then the following are true:*

(1) $\mathcal{F} = \sigma\text{-}alg < \cup_{n\geq 1}\mathcal{A}_n \cup \{F : \mu(F) = 0\} >$ *contains all Borel sets in Y.*
(2) *If $\mu(Y) < \infty$, $\alpha > 0$, f is a nonnegative and integrable function such that, for every n and every $J \in \mathcal{A}_n$ we have $J(f) \leq \alpha$, then we can conclude that $f(x) \leq \alpha$ almost everywhere.*

PROOF. (1) Suppose E is a closed subset of Y with positive measure. Since $\mathcal{A}_n$ is a partition of Y for every n, we can find $I_1^n, I_2^n, ... \in \mathcal{A}_n$ such that $\mu(I_j^n \cap E) > 0$ and $E = \bigcup_{j\geq 1}(I_j^n \cap E)$ *a.e.* Let $A_n = \bigcup_{j\geq 1} I_j^n$. We claim that $E = \bigcap_{n\geq 1} A_n$ *a.e.* To show this, suppose $x \in \bigcap_{n\geq 1} A_n$. This means that for every n and some j we have $x \in I_j^n$. Now we have

$$dist(x, E) \leq essdiam(I_j^n) \ a.e.$$

and this shows that almost everywhere $\bigcap_{n\geq 1} A_n \subset E$. Therefore $E \in \mathcal{F}$ and so is every Borel set.

(2). For every n define

$$f_n(x) = \sum J(f)\chi_J(x) \leq \alpha$$

where the summation is taken over all $J \in \mathcal{A}_n$. By defining $\mathcal{F}_n = \sigma\text{-}a\lg < \mathcal{A}_n >$ we have that $\mathcal{F}_1 \subset \mathcal{F}_2 \subset \dots$. Therefore f_n is a martingale relative to $\{\mathcal{F}_n \text{ , } n \geq 1\}$. Since

$$E(f_n) = \frac{1}{\mu(Y)}\int_Y f_n d\mu \leq \frac{1}{\mu(Y)}\int_Y \alpha d\mu \leq \alpha < \infty,$$

by the *Martingale Convergence Theorem* [**2**], we conclude that $\lim f_n(x)$ exists almost everywhere and converges to $f(x)$. Therefore $f(x) \leq \alpha$ almost everywhere. □

REMARK 4.2. Part (2) of the preceding lemma could be proved without using the Martingale Convergence Theorem. One proof is as follows: Let f_n be as in the lemma and define the linear operator $T_n : L^1(\mu) \longrightarrow L^1(\mu)$ by $T_n(f) = f - f_n$. It is easy to see that $T_n(f) \longrightarrow 0$ for every uniformly continuous function. Therefore $\{f \in L^1(\mu) : \|T_n(f)\| \longrightarrow 0\}$ is a closed linear subspace of $L^1(\mu)$ that contains every uniformly continuous function. Since the set of uniformly continuous functions is dense in $L^1(\mu)$ we conclude that $\|f - f_n\|_1 \longrightarrow 0$ for every $f \in L^1(\mu)$. Therefore $f_n \longrightarrow f$ a.e..

DEFINITION 4.3. In the application of the preceding lemma to a BMO triple $(X, \mu.\mathcal{I})$, we will insist that the partitions $\mathcal{A}_n$ are related to $\mathcal{I}$ (modulo sets of measure 0). Suppose B is a Borel subset of X, and $M > 1$. We say that B is *M-divisible* if there is a null sequence $\{\mathcal{A}_n\}$ of partitions of B such that

(1) $\mathcal{A}_0 = \{B\}$
(2) for every $n \geq 1$ and every $A \in \mathcal{A}_n$ and $C \in \mathcal{A}_{n+1}$, with $C \subset A$, we have $\mu(A) \leq M\mu(C)$
(3) for every $n \geq 1$ and every $A \in \mathcal{A}_n$ there is an $I \in \mathcal{I}$ such that $I \subset A$ and $\mu(A) \leq M\mu(I)$.

LEMMA 4.4. *Let $(X, \mu.\mathcal{I})$ be a BMO triple and suppose $I \in \mathcal{I}$ is M-divisible with respect to a null sequence $\{\mathcal{A}_n\}$ of partitions of I such that each $\mathcal{A}_n \subset \mathcal{I}$. Let $f \in L^1(I)$ be a* positive *function such that $I(f) < \alpha$. Then there is a finite or infinite sequence $\{I_j\}$ of disjoint subsets of I in $\mathcal{I}$ such that*

(1) *$f \leq \alpha$ almost everywhere on $I \backslash \cup_j I_j$,*
(2) *$\alpha \leq I_j(f) < M\alpha$,*
(3) *$\sum \mu(I_j) \leq \frac{1}{\alpha}\mu(I)I(f)$.*

PROOF. Let $\mathcal{E}_1 = \{J \in \mathcal{A}_1 : J(f) \geq \alpha\}$. And for each $n \geq 1$ let

$$\mathcal{E}_{n+1} = \mathcal{E}_n \cup \left\{J \in \mathcal{A}_{n+1} : J(f) \geq a \text{ and } J \cap \left(\bigcup_{A \in \mathcal{E}_n} A\right) = \varnothing\right\}$$

and let $\mathcal{E} = \bigcup_{n\geq 1} \mathcal{E}_n = \{I_1, I_2, \dots\}$.

Statement 1 follows from Lemma 4.1. For each $J \in \mathcal{E}$ there is an $n \geq 1$ and an $A \in \mathcal{A}_{n-1} \backslash \mathcal{E}$ such that $J \in \mathcal{E}_n$ and $J \subset A$. Then $J(f) \geq \alpha$ and $\alpha > A(f) \geq$

$\frac{1}{\mu(A)}\left[\mu\left(J\right)J\left(f\right)\right]$, which implies statement 2. Statement 3 follows immediately from statement 2. □

We now prove a general version of the *John-Nirenberg Theorem* [**7**]. Our proof is very close to the one in [**3**]

THEOREM 4.5. *Let $\varphi \in BMO(\mathcal{I},\mu)$ and let J be an M-divisible Borel subset of X. Then for every $\lambda > 0$ and every $\alpha \geq 1$,*

$$\mu\left\{t \in J : \left|\varphi(t) - J(\varphi)\right| > \lambda\right\} \leq \alpha\mu\left(J\right)\exp\left(\frac{-\ln\left(\alpha\right)\lambda}{6M^2\left\|\varphi\right\|_{BMO(\mathcal{I},\mu)}}\right).$$

PROOF. Note that if $\{\mathcal{A}_n\}$ is a null sequence of partitions of a Borel subset of X and if we replace $\mathcal{I}$ with $\mathcal{J} = \mathcal{I} \cup \bigcup_{n\geq 1}\mathcal{A}_n$, then it follows from Proposition 4 that $BMO(\mathcal{I},\mu) = BMO(\mathcal{J},\mu)$ and $\left\|\varphi\right\|_{BMO(\mathcal{J},\mu)} \leq 2M\left\|\varphi\right\|_{BMO(\mathcal{I},\mu)}$. This means that we can assume that $J \in \mathcal{I}$ and each $\mathcal{A}_n \subset \mathcal{I}$, but we have to prove the inequality with $4\alpha M^2$ replaced with $2\alpha M$.

Without loss of generality we can assume that $\left\|\varphi\right\|_{BMO(\mathcal{I},\mu)} = 1$. We apply Lemma 4.4 to $f = |\varphi(t) - J(\varphi)|$ to obtain $I_j^{(1)}$ as in Lemma 4.4. We have:

(1) $|\varphi(t) - J(\varphi)| \leq \alpha$, *a.e.* on $J \backslash \cup_j I_j^{(1)}$,
(2) $\left|I_j^{(1)}(\varphi) - J(\varphi)\right| < \alpha M$,
(3) $\sum \mu(I_j^{(1)}) \leq \frac{1}{\alpha}\mu(J)$.

On each $I_j^{(1)}$ we again apply Lemma 4.4 to $\left|\varphi - I_j^{(1)}(\varphi)\right|$ to obtain $I_j^{(2)}$ as before such that each $I_j^{(2)}$ is contained in some $I_j^{(1)}$. Then, *a.e.* on $J \backslash \cup_j I_j^{(2)}$, we will have:

$$\begin{aligned} |\varphi(t) - J(\varphi)| &\leq \left|\varphi(t) - I_j^{(1)}(\varphi)\right| + \left|I_j^{(1)}(\varphi) - J(\varphi)\right| \\ &< \alpha + \alpha M \ < 2\alpha M. \end{aligned}$$

Suppose $I_j^{(2)}$ is contained in $I_k^{(1)}$, then

$$\begin{aligned} \left|I_j^{(2)}(\varphi) - J(\varphi)\right| &\leq \left|I_j^{(2)}(\varphi) - I_k^{(1)}(\varphi)\right| + \left|I_k^{(1)}(\varphi) - J(\varphi)\right| \\ &< \alpha M + \alpha M = 2\alpha M. \end{aligned}$$

We also have:

$$\sum \mu(I_j^{(2)}) \leq \frac{1}{\alpha}\sum \mu(I_j^{(1)}) \leq (\frac{1}{\alpha})^2\mu(J).$$

Continue this process inductively. At stage n we get intervals $I_j^{(n)}$ such that

(1) $|\varphi(t) - J(\varphi)| \leq \alpha n M$, *a.e.* on $J \backslash \cup_j I_j^{(n)}$,
(2) $\sum \mu(I_j^{(n)}) \leq (\frac{1}{\alpha})^n\mu(J)$.

If $\alpha n M \leq \lambda < \alpha\left(n+1\right)M$, $n \geq 1$, then

$$\begin{aligned} \mu\left\{t \in J : |\varphi(t) - J(\varphi)| > \lambda\right\} &\leq \sum \mu(I_j^{(n)}) \\ &\leq (\frac{1}{\alpha})^n\mu(J) \leq e^{-c\lambda}\mu(J), \end{aligned}$$

for $c = \frac{1}{\alpha M} \ln \alpha$. Thus inequality in the theorem holds for $\alpha M \leq \lambda$. If $0 < \lambda < \alpha M$, then obviously

$$\mu \{t \in J : |\varphi(t) - J(\varphi)| > \lambda\} \leq \mu(J) < e^{\alpha M c} e^{-c\lambda} \mu(J).$$

Therefore the inequality holds for all λ. □

REMARK 4.6. (1) If every $I \in \mathcal{I}$ satisfies in the conditions of Lemma 4.4, then the previous theorem can be applied to show that for every $\varphi \in BMO(\mathcal{I}, \mu)$ and $p > 1$ there exists a constant A_p such that

$$\sup_{I \in \mathcal{I}} \left(\frac{1}{\mu(I)} \int_I | \varphi - I(\varphi) |^p \, d\mu \right)^{1/p} \leq A_p \|\varphi\|_{BMO(\mathcal{I},\mu)} .$$

(2) The converse of the John-Nirenberg theorem is also valid. In other words suppose φ is an integrable function on every $I \in \mathcal{I}$. If there are constants C and c such that $\forall I \in \mathcal{I}$, $\exists c_I \in \mathbb{C}$ such that

$$\mu \{t \in I : |\varphi(t) - c_I| > \lambda\} \leq C e^{-c\lambda} \mu(I)$$

for every $\lambda > 0$, then $\varphi \in BMO(\mathcal{I}, \mu)$.

(3) If $X = \mathbb{R}^2$, μ is *Lebesgue* measure, and $\mathcal{I}$ is the set of all disks, then every equilateral triangle is 4-divisible. We can partition a triangle into four triangles by joining the midpoints of the sides. With a little more work it can be shown that every disk is M-divisible for some $M > 1$.

(4) We could obtain more precision by choosing $\beta, M > 1$ and saying that a Borel set B is (M, β)-divisible if in Definition 4.3 we replace M with β in statement 3. In this case the right hand side of the John-Nirenberg inequality would replace M^2 with $M\beta$. In the triangle case in the preceding remark, we would get that every equilateral triangle would be $\left(4, \frac{3\sqrt{3}}{\pi}\right)$-divisible, so the $M^2 = 16$ could be replaced with $M\beta = 4\frac{3\sqrt{3}}{\pi} \approx 6.615\,9$.

(5) If X is a circle (interval) with $\mathcal{I}$ the set of open arcs (intervals), and if μ is any finite continuous (i.e., $\mu \{(x)\} = 0$ for every x) measure whose support is X, then every arc (interval) is 2-divisible; therefore the John-Niremberg theorem holds in $BMO(\mathcal{I}, \mu)$.

(6) If in our John-Nirenberg theorem we have $J \in \mathcal{I}$ and each $\mathcal{A}_n \subset \mathcal{I}$, then $4\alpha M^2$ can be replaced with $2\alpha M$ in the inequality.

5. Complements of VMO

The main result of this section is that the space $VMO(\mathcal{I}, \mu)$ is never complemented in $BMO(\mathcal{I}, \mu)$. The proof is based on a lemma that is adapted from [**4**].

LEMMA 5.1. *Suppose W is a normed space that has an uncountable subset $\mathcal{B}$ whose elements are linearly independent, and that there exists $M > 0$ such that for every $x_1, x_2, ..., x_n$ in $\mathcal{B}$ and every $\alpha_1, \alpha_2, ..., \alpha_n \in \mathbb{C}$,*

$$\left\| \sum_{k=1}^{n} \alpha_k x_k \right\| \leq M \max \{|\alpha_1|, |\alpha_2|, ..., |\alpha_n|\} .$$

Suppose also that Y is a topological vector space with continuous linear functionals $\varphi_1, \varphi_2, ... : Y \longrightarrow \mathbb{C}$, that separate the points of Y. Then there is no injective continuous linear map $f : W \longrightarrow Y$.

PROOF. Suppose, via contradiction, that a map f exists. For every n the map $\varphi_n \circ f$ is a bounded linear functional on W. Let $E_{n,k} = \{x \in \mathcal{B} : |\varphi_n(f(x))| \geq \frac{1}{k}\}$. Since the function f is 1-1 and the elements of $\mathcal{B}$ are linearly independent, then $\mathcal{B} = \bigcup_{k,n} E_{n,k}$. Thus there exist n_0 and k_0 such that E_{n_0,k_0} is uncountable. Choose distinct elements $x_1, x_2, ... \in E_{n_0,k_0}$ and, for the sake of simplicity, define $\varphi_{n_0}(f(x_k)) = r_k\, e^{i\theta_k}$, $x = \sum_{k=1}^{n} e^{-i\theta_k} x_k$. Then $\|x\| \leq M$ and for every n we have:

$$M \left\|\varphi_{n_0} \circ f\right\| \geq \|x\| \left\|\varphi_{n_0} \circ f\right\| \geq \left|\varphi_{n_0}(f(x))\right| = \sum_{k=1}^{n} r_k \geq \frac{n}{k_0},$$

which is a contradiction. □

THEOREM 5.2. *There is no injective continuous linear map*

$$\varphi : BMO(\mathcal{I},\mu)/VMO(\mathcal{I},\mu) \longrightarrow BMO(\mathcal{I},\mu).$$

In particular, $VMO(\mathcal{I},\mu)$ is not complemented in $BMO(\mathcal{I},\mu)$.

PROOF. By Lemma 2.4 the points of $BMO(\mathcal{I},\mu)$ are separated by countably many continuous linear functionals. By Lemma 5.1 it is enough to find uncountably many functions on $BMO(\mathcal{I},\mu)/VMO(\mathcal{I},\mu)$ that are linearly independent and that satisfy in an inequality as in Lemma 5.1. To do so, suppose $x \in X$. By using the second property of $\mathcal{I}$, choose I_n in $\overline{\mathcal{B}(x;\frac{1}{n})\backslash\mathcal{B}(x;\frac{1}{n+1})}$, and, by the regularity of μ, choose compact subsets A_n and B_n of I_n so that

$$\mu(A_n) \approx \mu(B_n) \approx \frac{1}{2}\mu(I_n).$$

Since I_n "converges to" $\{x\}$, the sets $A = \cup_{n\geq 1} A_n$ and $B = \cup_{n\geq 1} B_n$ are disjoint closed subsets of the space $X\backslash\{x\}$. Define p_x on X by

$$p_x(x) = 0, \text{ and } p_x(y) = \frac{d(y,A)}{d(y,A) + d(y,B)} \quad \forall y \in X\backslash\{x\}.$$

Then the function p_x is bounded by 1 (and so belongs to $BMO(\mathcal{I},\mu)$), $p_x|_A = 0$, and $p_x|_B = 1$ (and so $p_x \notin VMO(\mathcal{I},\mu)$). Thus p_x is a nonzero function in the quotient space $BMO(\mathcal{I},\mu)/VMO(\mathcal{I},\mu)$. It is also easy to see that the function p_x is uniformly continuous on $X\backslash\mathcal{B}(x;\varepsilon)$ for every $\varepsilon > 0$. The set

$$\mathcal{B} = \{p_x : x \in X\}$$

is an uncountable subset of $BMO(\mathcal{I},\mu)/VMO(\mathcal{I},\mu)$ whose elements are linearly independent. By Lemma 2.4, every uniformly continuous function is in $VMO(\mathcal{I},\mu)$ and so p_x, as a function in $BMO(\mathcal{I},\mu)/VMO(\mathcal{I},\mu)$, is zero everywhere except on $\mathcal{B}(x;\epsilon)$ for every $\epsilon > 0$.This fact can be used to show that:

$$\left\|\sum_{k=1}^{n} \alpha_k\, p_{x_k}\right\|_{BMO(\mathcal{I},\mu)/VMO(\mathcal{I},\mu)} \leq 3\max\{|\alpha_1|, |\alpha_2|, ..., |\alpha_n|\},$$

for every $p_{x_1}, p_{x_2}, ..., p_{x_n}$ in $\mathcal{B}$ and every $\alpha_1, \alpha_2, ..., \alpha_n \in \mathbb{C}$. This completes the proof. □

The following Corollary follows from Proposition 2.

COROLLARY 3. *$C_u(X) \cap L^\infty(\mu)$ is not complemented in $L^\infty(\mu)$.*

References

[1] R. Coifman and G. Weiss, Extensions of Hardy Spaces and their uses in Analysis, Bull. Amer. Math. Soc., 83 (1977), 569-645.
[2] J. Doob, What is a Martingale?, Amer. Math. Monthly 78 (1971), 451–463.
[3] J. Garnett, Bounded Analytic Functions, Academic Press INC., 1981.
[4] L. Ge, D. Hadwin, Ultraproducts of C^*-algebras, Operator Theory: Advances and Applications, 127 (2001), 305-326.
[5] D. Hadwin, Continuity Modulo Sets of Measure Zero, Mathematica Balkanica, Vol. 3 (1989), 430-433.
[6] F. John, Rotation and Strain, Comm. Pure Appl. Math. 14 (1961), 391-413.
[7] F. John and L. Nirenberg, On Function of Bounded Mean Oscillation, Comm. Pure Appl. Math. 14 (1961), 415-426.
[8] K. Peterson, Brownian Motion, Hardy Spaces and Bounded Mean Oscillation, Cambridge Univ. Press, Cambridge, 1977.
[9] D. Sarason, Functions of Vanishing Mean Oscillation, Trans. Amer. Math. Soc. 207 (1975), 391-405.
[10] D. Sarason, Function Theory on the Unit Circle, Virginia Poly. Inst. and State Univ., Blacksburg 1978.

Mathematics Department, University of New Hampshire
E-mail address: don@unh.edu
URL: http://www.math.unh.edu/~don

Mathematics Department, California State University Fullerton
E-mail address: hyousefi@fullerton.edu

Contemporary Mathematics
Volume **454**, 2008

Order Bounded Weighted Composition Operators

R. A. Hibschweiler

Abstract. Let X be a Banach space of functions analytic in the unit disc and let m denote normalized Lebesgue measure on the circle. The operator $T : X \to L^q(m)$ is said to be order-bounded if there exists $h \in L^q(m)$ such that $| (Tf)(e^{i\theta}) | \leq h(e^{i\theta})$ a.e. $[m]$ for all $\| f \|_X \leq 1$. Let $\Psi \neq 0 \in L^q(m)$ and let Φ be an analytic self-map of the disc. The weighted composition operator $W_{\Psi,\Phi}$ is defined by $W_{\Psi,\Phi}(f) = \Psi(f \circ \Phi)$ for functions f analytic in the disc. Order boundedness of $W_{\Psi,\Phi}$ is studied on the weighted Bergman spaces and on more general Banach spaces of analytic functions with restricted growth. Connections are exposed between boundedness, compactness and order boundedness on the weighted Bergman spaces.

1. Introduction

For $p \geq 1$, the Hardy space H^p is the Banach space of functions analytic in the disc such that

$$\| f \|_{H^p}^p = \frac{1}{2\pi} \sup_{0 \leq r < 1} \int_0^{2\pi} | f(re^{i\theta}) |^p \ d\theta < \infty.$$

Recall that if $f \in H^p$, then the boundary function f^* defined by

$$f^*(e^{i\theta}) = \lim_{r \to 1-} f(re^{i\theta})$$

exists a.e. with respect to Lebesgue measure m. Let D denote the open unit disc. Throughout this work, Φ will denote an analytic function mapping D into itself. As is well known, the composition operator defined by $C_\Phi(f) = f \circ \Phi$ is bounded on H^p for each $p \geq 1$.

The focus of this work is on order bounded weighted composition operators acting on the Hardy spaces and the weighted Bergman spaces. The general definition of order bounded operators is given next.

Definition 1.1. Let X be a Banach space of functions analytic in D and let $q > 0$. Let μ be a positive measure on the unit circle. The operator $T : X \to L^q(\mu)$ is said to be order bounded if there exists $h \in L^q(\mu), h \geq 0$, so that the inequality

$$| T(f)(e^{i\theta}) | \leq \ h(e^{i\theta})$$

1991 *Mathematics Subject Classification.* Primary 54C40, 14E20; Secondary 46E25, 20C20.
Key words and phrases. weighted composition operators, weighted Bergman spaces.

holds a.e. with respect to μ, for all $f \in X$ with $\| f \|_X \leq 1$.

In particular, let $p \geq 1$ and $\beta > 0$ and let Φ be an analytic self-map of D such that $\Phi^* \in L^{p\beta}(m)$. H. Hunziker [**5**] characterized the self-maps Φ for which the composition operator $C_\Phi : H^p \to L^{\beta p}(m)$ is order bounded. In this context, $C_\Phi(f)$ is understood to mean the boundary function $(f \circ \Phi)^*$.

THEOREM 1.2 (Hunziker). *Let $\beta > 0$. The following are equivalent.*

(1) *$C_\Phi : H^p \to L^{\beta p}(m)$ is order bounded for some $p \geq 1$.*
(2) *$1/(1- \mid \Phi^* \mid) \in L^\beta(m)$.*

Since the condition (2) in Hunziker's theorem is independent of p, it is clear that if $C_\Phi : H^p \to L^{\beta p}(m)$ is order bounded for some $p \geq 1$, then it is order bounded for all such p.

Assume $p > 0$ and $q > 0$. Let $0 \neq \Psi \in L^q(m)$. The weighted composition operator $W_{\Psi,\Phi}$ is defined by $W_{\Psi,\Phi}(f) = \Psi\,(f \circ \Phi)$ where f is analytic in the disc. The focus of this work is on weighted composition operators on the Hardy spaces, and more generally, on the weighted Bergman spaces. Definition 1.1 implies that if $W_{\Psi,\Phi} : H^p \to L^q(m)$ is order bounded, then $W_{\Psi,\Phi} : H^p \to L^q(m)$ is bounded. The converse is false, as shown by Hunziker's theorem and the simple example $p = q$, $\Psi = 1$, and $\Phi(z) = z$.

Let $dA(z)$ denote normalized area measure on the disc. For $\alpha > -1$ and $p \geq 1$, the weighted Bergman space A^p_α is the set of functions analytic in the disc satisfying

$$\| f \|^p_{A^p_\alpha} = \int_D \mid f(z) \mid^p \; (1- \mid z \mid^2)^\alpha \, dA(z) \; < \infty.$$

Also note that for $\alpha = -1$, the appropriate definition for A^p_α is the Hardy space H^p [**14**].

In contrast with the Hardy spaces, the Bergman spaces include functions that have no boundary values. See, for example, [**3**]. Thus a discussion of order bounded weighted composition operators on the Bergman spaces will require the assumption that $\mid \Phi^*(e^{i\theta}) \mid < 1$ a. e. with respect to m. Hunziker's theorem indicates that this condition is necessary for $C_\Phi : H^p \to L^{\beta p}(m)$ to be order bounded.

Let $z \in D$ and let $E_z(f) = f(z)$ for $f \in A^p_\alpha$. The following lemma is well known, and so only a sketch of the proof will be given here.

LEMMA 1.3. *Fix $\alpha \geq -1$ and $p \geq 1$. There are positive constants C_1 and C_2 depending only on α and p such that*

$$C_1(1- \mid z \mid)^{-(\alpha+2)/p} \; \leq \| E_z \| \leq \; C_2(1- \mid z \mid)^{-(\alpha+2)/p}.$$

PROOF. For $\alpha = -1$, see [**2**] and Proposition 1.1 [**6**]. For $\alpha > -1$, Smith [**14**] used the subharmonicity of $\mid f \mid^p$ to establish

$$\mid f(z) \mid \leq \; C_2 \, \| f \|_{A^p_\alpha} \; (1- \mid z \mid)^{-(\alpha+2)/p}$$

where C_2 depends only on α and p, and f is any function in A^p_α. This yields the second inequality in the lemma. For the remaining inequality, let

$$f_z(w) = \frac{(1- \mid z \mid^2)^{(\alpha+2)/p}}{(1 - \overline{z}w)^{2(\alpha+2)/p}}, \mid w \mid < 1.$$

Then $\| f_z \|_{A^p_\alpha} \approx 1$ [**14**], and thus there are constants K_1 and K_2 depending only on α and p such that $K_1 \leq \| f_z \|_{A^p_\alpha} \leq K_2$. Therefore

$$\| E_z \| \geq \frac{| f_z(z) |}{\| f_z \|_{A^p_\alpha}} \geq \frac{1}{K_2} (1 - | z |^2)^{-(\alpha+2)/p}.$$

The proof is complete. □

In what follows, C denotes a generic positive constant.

THEOREM 1.4. *Let $\alpha \geq -1$, $q > 0$ and $\Psi \in L^q(m)$. Let Φ be an analytic self-map of the disc such that $| \Phi(e^{i\theta}) | < 1$ a.e $[m]$. For fixed p, $1 \leq p < \infty$, the following are equivalent.*

(1) *$W_{\Psi,\Phi} : A^p_\alpha \to L^q(m)$ is order bounded.*
(2) *$\Psi/(1 - | \Phi^* |)^{(\alpha+2)/p} \in L^q(m)$.*

PROOF. Suppose that $\Psi/(1 - | \Phi^* |)^{(\alpha+2)/p} \in L^q(m)$. Since $| \Phi^*(e^{i\theta}) | < 1$ a.e. $[m]$, Lemma 1.3 provides a constant C depending only on α and p such that

$$| f(\Phi^*(e^{i\theta})) | \leq C(1 - | \Phi^*(e^{i\theta}) |)^{-(\alpha+2)/p}$$

a.e. $[m]$ for all f with $\| f \|_{A^p_\alpha} \leq 1$. Let

$$h(e^{i\theta}) = C | \Psi(e^{i\theta}) | (1 - | \Phi^*(e^{i\theta}) |)^{-(\alpha+2)/p}.$$

Then $h \in L^q(m)$, by hypothesis, and the previous inequality implies that

$$| \Psi(e^{i\theta}) | \, | f(\Phi^*(e^{i\theta})) | \leq h(e^{i\theta})$$

a.e. $[m]$. Thus $W_{\Psi,\Phi} : A^p_\alpha \to L^q(m)$ is order bounded, as required.

Next suppose that $W_{\Psi,\Phi} : A^p_\alpha \to L^q(m)$ is order bounded. Thus there exists $h \in L^q(m), h \geq 0$, with

$$| h(e^{i\theta}) | \geq | \Psi(e^{i\theta}) | \, | f(\Phi^*(e^{i\theta})) |$$

a.e. $[m]$ for every f with $\| f \|_{A^p_\alpha} \leq 1$. By Lemma 1.3, the inequality

$$h(e^{i\theta}) \geq | \Psi(e^{i\theta}) | \sup\{| E_{\Phi^*(e^{i\theta})}(f) | : \| f \|_{A^p_\alpha} \leq 1\}$$

$$= | \Psi(e^{i\theta}) | \, \| E_{\Phi^*(e^{i\theta})} \| \geq C | \Psi(e^{i\theta}) | (1 - | \Phi^*(e^{i\theta}) |)^{-(\alpha+2)/p}$$

holds a.e. $[m]$. It follows that $\Psi/(1 - | \Phi^* |)^{-(\alpha+2)/p} \in L^q(m)$. □

COROLLARY 1.5. *Fix $\alpha \geq -1$, $p \geq 1$ and $q > 0$. Let n be a natural number. The following are equivalent.*

(1) *$W_{\Psi,\Phi} : A^p_\alpha \to L^q(m)$ is order bounded.*
(2) *$W_{\Psi,\Phi^n} : A^p_\alpha \to L^q(m)$ is order bounded.*

PROOF. Since $1 - | \Phi^*(e^{i\theta}) | \leq 1 - | (\Phi^*)^n(e^{i\theta}) | \leq n(1 - | \Phi^*(e^{i\theta}) |)$, Theorem 1.4 gives the result. □

Let $\alpha \geq -1$ and $p, q > 0$. T. Domenig [**1**] proved that $C_\Phi : A^p_\alpha \to L^q(m)$ is order bounded if and only if $1/(1 - | \Phi^* |)^{(\alpha+2)/p} \in L^q(m)$. His result is recovered here as the case $\Psi = 1$. As a consequence of Theorem 1.1 (Hunziker) and Domenig's theorem, $C_\Phi : A^p_\alpha \to L^q(m)$ is order bounded for fixed $\alpha > -1$ if and only if $C_\Phi : H^p \to L^{(\alpha+2)q}(m)$ is order bounded. A version of this result is possible for the weighted composition operator $W_{\Psi,\Phi}$. Suppose that $\Psi \in L^\infty$ and $W_{\Psi,\Phi} : A^p_\alpha \to$

$L^q(m)$ is order bounded for some $\alpha > -1$. By Theorem 1.4, $\Psi/(1-|\Phi^*|)^{(\alpha+2)/p} \in L^q(m)$. It follows that

$$\int_0^{2\pi} \left(\frac{|\Psi|}{(1-|\Phi^*|)^{1/p}}\right)^{(\alpha+2)q} dm < \|\Psi\|_\infty^{(\alpha+1)q} \int_0^{2\pi} \frac{|\Psi|^q}{(1-|\Phi^*|)^{(\alpha+2)q/p}} dm < \infty.$$

Thus $\Psi/(1-|\Phi^*|)^{1/p} \in L^{(\alpha+2)q}(m)$. By Theorem 1.4, $W_{\Psi,\Phi} : H^p \to L^{(\alpha+2)q}(m)$ is order bounded.

Suppose that $W_{\Psi,\Phi} : H^p \to L^{(\beta+2)q}(m)$ is order bounded for some $\beta > -1$ and Ψ is bounded away from 0, that is, there is a positive constant C such that $C \leq |\Psi(e^{i\theta})|$ a.e. $[m]$. An argument using Theorem 1.4 implies that $W_{\Psi,\Phi} : A^p_\beta \to L^q(m)$ is order bounded. The details are omitted.

2. Weighted Dirichlet Spaces

For $\gamma > -1$, the weighted Dirichlet space D_γ is the Hilbert space of analytic functions $f = \sum_{n=0}^\infty a_n z^n$, $(|z| < 1)$ with

$$\|f\|^2_{D_\gamma} = \sum_{n=0}^\infty \frac{|a_n|^2}{(n+1)^{\gamma-1}} < \infty.$$

The functions $e_{\gamma,n} = (n+1)^{(\gamma-1)/2} z^n$, $n = 0, 1, 2, \ldots$ are an orthonormal basis for D_γ. Note that D_1 is the Hardy space H^2 and $D_\gamma = A^2_{\gamma-2}$ for $\gamma > 1$.

The operator $T : D_\gamma \to H^2$ is Hilbert-Schmidt if and only if

$$\sum_{n=0}^\infty \|T(e_{\gamma,n})\|^2_{H^2} < \infty.$$

In [**13**], J. H. Shapiro and P. Taylor proved that $C_\Phi : H^2 \to H^2$ is Hilbert-Schmidt if and only if $1/(1-|\Phi^*|) \in L^1(m)$. H. Jarchow and R. Riedl [**7**] proved that for $\beta > 0$, $C_\Phi : D_\beta \to H^2$ is Hilbert-Schmidt if and only if $C_\Phi : H^p \to L^{p\beta}(m)$ is order bounded for every $p \geq 1$. These ideas will be expanded here to the setting of the weighted Bergman spaces. In the rest of this section, Φ will denote an analytic self-map of D such that $|\Phi^*(e^{i\theta})| < 1$ a. e. $[m]$.

THEOREM 2.1. *Let $\alpha \geq -1$, $\beta > 0$ and $\gamma = (\alpha+2)\beta$. The following are equivalent.*

(1) *$C_\Phi : A^p_\alpha \to L^{p\beta}(m)$ is order bounded for some (for all) $p > 0$.*
(2) *$C_\Phi : D_\gamma \to H^2$ is Hilbert-Schmidt.*

PROOF. Because of Jarchow and Riedl's result, it is enough to prove the corollary in case $\alpha > -1$. Suppose that $C_\Phi : A^p_\alpha \to L^{p\beta}(m)$ is order bounded. Domenig's theorem yields $1/(1-|\Phi^*|) \in L^{(\alpha+2)\beta}(m)$. Hunziker's theorem now yields that $C_\Phi : H^p \to L^{(\alpha+2)p\beta}(m)$ is order bounded. Therefore $C_\Phi : D_\gamma \to H^2$ is Hilbert-Schmidt. These steps can be reversed to prove the remaining implication. □

If $\Psi \in L^\infty(m)$ and if there is a positive constant c such that the inequality $c \leq |\Psi|$ holds a.e. $[m]$, then a result analogous to the previous corollary holds for the operator $W_{\Psi,\Phi}$. The statement is omitted.

Fix $\gamma > 0$ and let $(1-z)^{-\gamma} = \sum_{n=0}^\infty A_n(\gamma) z^n$, $|z| < 1$. By Stirling's formula, $A_n(\gamma) \approx (n+1)^{\gamma-1}$ as $n \to \infty$.

THEOREM 2.2. *Suppose that $k \in \mathbf{N}, \alpha \geq -1$ and $\Psi \in L^{2k}(m)$. Fix $p \geq 1$ and let $\gamma = 2k(\alpha+2)/p$. The following are equivalent.*

(1) $W_{\Psi,\Phi} : A^p_\alpha \to L^{2k}(m)$ *is order bounded.*
(2) $W_{\Psi^k,\Phi} : D_\gamma \to H^2$ *is Hilbert-Schmidt.*

PROOF. By Theorem 1.4, $W_{\Psi,\Phi} : A^p_\alpha \to L^{2k}(m)$ is order bounded

$$\Leftrightarrow \frac{\Psi}{(1- \mid \Phi^* \mid^2)^{(\alpha+2)/p}} \in L^{2k}(m)$$

$$\Leftrightarrow \int_0^{2\pi} \mid \Psi \mid^{2k} \sum_{n=0}^{\infty} A_n(\gamma) \mid \Phi^* \mid^{2n} \, dm < \infty$$

$$\Leftrightarrow \sum_{n=0}^{\infty} \| W_{\Psi^k,\Phi}(e_{\gamma,n}) \|^2_{H^2} < \infty$$

$$\Leftrightarrow W_{\Psi^k,\Phi} : D_\gamma \to H^2 \text{ is Hilbert-Schmidt.}$$

□

3. Weighted Banach Spaces

In this section a connection is drawn between β-order boundedness of $W_{\Psi,\Phi}$ on the Bergman spaces and on certain Banach spaces defined through the use of a weight function v.

DEFINITION 3.1. A weight is a non-increasing, continuous function

$$v : [0,1] \to \mathbf{R}$$

with the properties $v(r) > 0$ for $0 \le r < 1$ and $v(1) = 0$.

For $z \in D$, the notation $v(z)$ will be used to denote $v(\mid z \mid)$.

DEFINITION 3.2.

$$H^\infty_v \;=\; \{f \in H(D) : \| f \|_v = \sup_{z \in D} \mid f(z) \mid v(z) \; < \infty\}.$$

For any weight v, H^∞_v is a Banach space. In what follows we will be interested in weights of the form $v(r) = (1-r)^k$, $k > 0$. A more general version of the following result is due to A. Montes-Rodriguez [**11**].

LEMMA 3.3. *Fix $k > 0$ and let $v(r) = (1-r)^k$. For $z \in D$,let $E_z(f) = f(z)$ for $f \in H^\infty_v$. Then*

$$2^{-k} \, (1- \mid z \mid)^{-k} \;\le \| E_z \| \le \; (1- \mid z \mid)^{-k}.$$

PROOF. Fix $z \in D$. Then

$$\mid E_z(f) \mid = \frac{\mid f(z) \mid \; v(z)}{v(z)} \;\le \frac{\| f \|_v}{v(z)}.$$

For the remaining inequality, consider the function $f_0(w) = (1-\bar{z}w)^{-k}$ $(w \in D)$. □

Theorem 3.5 will establish a connection between order boundedness and boundedness on the spaces H^∞_v. The following lemma is needed in the proof.

LEMMA 3.4. *For $k > 0$ and $0 \le r < 1$, there is a positive constant C depending only on k such that*

$$\frac{r^2}{(1-r^2)^{2k}} \;\le\; C \sum_{n=0}^{\infty} 2^{2nk} r^{2^{n+1}}.$$

PROOF. Let I denote the sum in the previous expression. For $n = 0, 1, 2, \ldots$ let $I_n = \{m \in \mathbf{Z} : 2^n - 1 \leq m < 2^{n+1} - 1\}$. Since there are exactly 2^n terms in I_n, it follows that

$$I = \sum_{n=0}^{\infty} 2^{-n} \Big(\sum_{m \in I_n} 2^{2nk} \Big) r^{2^{n+1}}.$$

If $m \in I_n$, then $(m+1)/2 < 2^n \leq m + 1$. It follows that

$$I \geq 2^{-2k} \sum_{n=0}^{\infty} \Big(\sum_{m \in I_n} (m+1)^{2k-1} r^{2m+2} \Big) \text{ for } 0 \leq r < 1.$$

Stirling's formula now implies that

$$I \geq C\, r^2\, 2^{-2k} \sum_{n=0}^{\infty} A_n(2k)\, (r^2)^n = C \frac{r^2}{(1-r^2)^{2k}}.$$

□

Recall the assumption that $| \Phi^*(e^{i\theta}) | < 1$ a. e. $[m]$.

The proof of Theorem 3.5 will use Khinchine's inequality. A statement can be found in Luecking's paper [**8**].

THEOREM 3.5. *Fix $k, q > 0$ and let $v(r) = (1-r)^k$. Let $0 \neq \Psi \in L^q(m)$. The following are equivalent.*

(1) *$W_{\Psi,\Phi} : H_v^\infty \to L^q(m)$ is order bounded.*
(2) *$W_{\Psi,\Phi} : H_v^\infty \to L^q(m)$ is bounded.*
(3) *$\Psi/(v \circ \Phi^*) \in L^q(m)$.*

PROOF. It will be shown that (1) $\Rightarrow$ (2) $\Rightarrow$ (3) $\Rightarrow$ (1).

First assume that $W_{\Psi,\Phi} : H_v^\infty \to L^q(m)$ is order bounded. Thus there exists a positive function $h \in L^q(m)$ such that

$$| \Psi(e^{i\theta})\, f(\Phi^*(e^{i\theta})) | \leq h(e^{i\theta}) \text{ a.e. } [m]$$

for all $f \in H_v^\infty$ with $\| f \|_v \leq 1$. It now follows that

$$\| \Psi(f \circ \Phi^*) \|_{L^q(m)} \leq \| h \|_{L^q(m)}\, \| f \|_v$$

for $f \in H_v^\infty$.

Next assume that $W_{\Psi,\Phi} : H_v^\infty \to L^q(m)$ is bounded. Thus there is a positive constant K such that

$$\| W_{\Psi,\Phi}(f) \|_{L^q(m)} \leq K \; \| f \|_v$$

for all $f \in H_v^\infty$. Let

$$f(z) = \sum_{n=1}^{\infty} 2^{kn} z^{2^n}, | z | < 1.$$

In [**7**], Jarchow and Riedl showed that $f \in H_v^\infty$ in the case $0 < k \leq 1$. However, their argument remains valid for all $k > 0$. Let $r_n(t)$ denote the Rademacher functions given by

$$r_0(t) = \begin{cases} 1, & \text{if } 0 \leq t - [t] < 1/2, \\ -1, & \text{if } 1/2 \leq t - [t] < 1. \end{cases}$$

For $n \geq 1$, let $r_n(t) = r_0(2^n t)$. For $0 \leq t \leq 1$,

$$f_t(z) = \sum_{n=1}^{\infty} 2^{kn}\ r_n(t)\ z^{2^n}\ (\mid z \mid < 1).$$

Since Jarchow and Riedl's estimate on $\| f \|_v$ depends solely on the magnitude of the coefficients of f and since $\mid r_n(t) \mid = 1$ for all n and t, there is a constant C such that $\| f_t \|_v \leq C$ for $0 \leq t \leq 1$. It follows that $\| W_{\Psi,\Phi}(f_t) \|_{L^q(m)} \leq KC$ for $0 \leq t \leq 1$. Thus

$$(KC)^q \geq \int_0^{2\pi} \mid \Psi(e^{i\theta}) \mid^q \mid \sum_{n=0}^{\infty} 2^{kn} r_n(t)\ \Phi^*(e^{i\theta})^{2^n} \mid^q\ dm \text{ for } 0 \leq t \leq 1.$$

Since the previous estimate holds for all $0 \leq t \leq 1$, integration with respect to t and Fubini's theorem yield

$$(KC)^q \geq \int_0^{2\pi} \mid \Psi(e^{i\theta}) \mid^q \int_0^1 \mid \sum_0^{\infty} 2^{kn}\ r_n(t)\ \Phi^*(e^{i\theta})^{2^n} \mid^q\ dt\ dm.$$

Khinchine's inequality provides a lower bound for the inner integral in the previous expression, and yields

$$C_1 \geq \int_0^{2\pi} \mid \Psi(e^{i\theta}) \mid^q \ (\sum_0^{\infty} 2^{2nk} \mid \Phi^*(e^{i\theta}) \mid^{2^{n+1}})^{q/2}\ dm$$

where C_1 is a positive constant depending only on q. Bringing all the constants together, Lemma 3.4 now implies that there is a positive constant C_0 such that

$$C_0 \geq \int_0^{2\pi} \mid \Psi(e^{i\theta}) \mid^q \ \frac{\mid \Phi^*(e^{i\theta}) \mid^q}{(1- \mid \Phi^*(e^{i\theta}) \mid^2)^{kq}}\ dm.$$

An easy argument now yields $\Psi/(v \circ \Phi^*) \in L^q(m)$.

Finally suppose that $\Psi/(v \circ \Phi^*) \in L^q(m)$ and let $f \in H_v^\infty$ with $\| f \|_v \leq 1$. Since $\mid \Phi^*(e^{i\theta}) \mid < 1$ a.e $[m]$, Lemma 3.3 implies that the inequality

$$\mid W_{\Psi,\Phi}(f)(e^{i\theta}) \mid \leq \mid \Psi(e^{i\theta}) \mid \| E_{\Phi^*(e^{i\theta})} \| \| f \|_v \leq \mid \Psi(e^{i\theta}) \mid \ \frac{1}{v(\Phi^*(e^{i\theta}))}$$

holds a.e. $[m]$. Thus the function $h = \mid \Psi \mid \ (1- \mid \Phi^* \mid)^{-k}$ serves as the dominating function in Definition 1.1, and $W_{\Psi,\Phi} : H_v^\infty \to L^q(m)$ is order bounded. The proof is complete. □

Let $w_p(r) = (1-r)^{1/p}$. Lemma 1.3 implies that $H^p \subset H^\infty_{w_p}$. Let $\beta > 0$. In [**7**], Jarchow and Riedl proved that $C_\Phi : H^p \to L^{p\beta}(m)$ is order bounded for some (for all) $p \geq 1$ if and only if $C_\Phi : H_{w_p} \to L^{p\beta}(m)$ is order bounded for some (for all) $p \geq 1$. The next corollaries present related facts for $W_{\Psi,\Phi}$ in the context of the weighted Bergman spaces. Here the appropriate weight will be $v_{\alpha,p}(r) = (1-r)^{(\alpha+2)/p}$, where $p \geq 1$ and $\alpha \geq -1$. Lemma 1.3 shows that for each such p, $A^p_\alpha \subset H^\infty_{v_{\alpha,p}}$. Corollary 3.6 follows from Theorem 1.4 and Theorem 3.5.

COROLLARY 3.6. *Fix $p \geq 1, \alpha \geq -1$ and let $\beta > 0$. The following are equivalent.*

(1) *$W_{\Psi,\Phi} : A^p_\alpha \to L^{p\beta}(m)$ is order bounded.*
(2) *$W_{\Psi,\Phi} : H^\infty_{v_{\alpha,p}} \to L^{p\beta}(m)$ is order bounded.*

Putting $\Psi = 1$ in Corollary 3.7 yields the Bergman space analogue of Jarchow and Riedl's result, mentioned above. The proof of the corollary is omitted.

COROLLARY 3.7. *Fix $\alpha \geq -1$ and $\beta > 0$. Suppose that $\Psi \in L^\infty(m)$ and there is a constant c such that $c \leq |\Psi(e^{i\theta})|$ a.e. $[m]$. The following are equivalent.*

(1) *$W_{\Psi,\Phi} : H^\infty_{v_{\alpha,p}} \to L^{p\beta}(m)$ is order bounded for some (for all) $p \geq 1$.*
(2) *$W_{\Psi,\Phi} : A^p_\alpha \to L^{p\beta}(m)$ is order bounded for some (for all) $p \geq 1$.*

4. A Characterization of Boundedness and Compactness

In [**6**], Hunziker and Jarchow found relationships between order boundedness, boundedness and compactness of the operator C_Φ on the Hardy spaces. Analogous results are given here for $W_{\Psi,\Phi}$ on the Bergman spaces.

THEOREM 4.1 (Hunziker and Jarchow). (1) *If $\beta \geq 1$ and $C_\Phi : H^p \to L^{p\beta}(m)$ is order bounded for some $p \geq 1$, then $C_\Phi : H^p \to H^{p\beta}$ is compact for all $p \geq 1$.*
(2) *If $C_\Phi : H^p \to H^{p\gamma}$ is bounded for some $p \geq 1$ and $0 < \beta < \gamma - 1$, then $C_\Phi : H^p \to L^{p\beta}(m)$ is order bounded for all $p \geq 1$.*

The converse of assertion (1) is false. To see this in the case $\beta = 1$, note that J. H. Shapiro and P. Taylor proved that there are compact composition operators $C_\Phi : H^2 \to H^2$ which are not Hilbert-Schmidt [**13**]. By Theorem 3.1 [**13**], it follows that $1/(1-|\Phi^*|) \notin L^1(m)$. By Hunziker's result, stated here as Theorem 1.2, $C_\Phi : H^p \to L^p(m)$ is not order bounded.

In Corollaries 4.2, 4.3 and 4.4, recall the assumption that Φ is a self-map of D with $|\Phi^*| < 1$ a. e. $[m]$.

COROLLARY 4.2. *Suppose that $\alpha \geq -1$, $\beta > 0$ and $C_\Phi : A^p_\alpha \to L^{p\beta}(m)$ is order bounded. If $(\alpha+2)\beta \geq 1$, then $C_\Phi : H^p \to H^{p(\alpha+2)\beta}$ is compact for all $p \geq 1$.*

PROOF. Since $C_\Phi : A^p_\alpha \to L^{p\beta}(m)$ is order bounded, Domenig's theorem [**1**] yields $1/(1-|\Phi^*|) \in L^{(\alpha+2)\beta}(m)$. Hunziker's Theorem (Theorem 1.2) implies that $C_\Phi : H^p \to L^{p(\alpha+2)\beta}(m)$ is order bounded for all $p \geq 1$. The result now follows by Theorem 4.1 (Part 1). □

R. Riedl [**12**] used the classical Nevanlinna counting function to characterize self-maps Φ which induce bounded or compact composition operators $C_\Phi : H^p \to H^q$ in the case $0 < p \leq q$. In [**14**], W. Smith used the generalized Nevanlinna counting function to characterize bounded or compact composition operators $C_\Phi : A^p_\alpha \to A^q_\beta$ in the case $0 < p \leq q$. These results will expose further connections between order boundedness, boundedness and compactness.

For Φ a self-map of D, $w \neq \Phi(0)$ and $\gamma > 0$,

$$N_\gamma(w) = \sum (\log(1/|z|))^\gamma$$

where the sum extends over all z with $\Phi(z) = w$, counting multiplicities. Thus the classical Nevanlinna counting function is $N_1(w)$. Riedl [**12**] proved that $C_\Phi : H^p \to H^q$ is bounded in the case $0 < p \leq q$ if and only if

$$N_1(w) = O((1-|w|)^{q/p}), \ |w| \to 1.$$

Smith [**14**] showed that for $0 < p \leq q$, $C_\Phi : A^p_\alpha \to A^q_\beta$ is bounded if and only if

$$N_{\beta+2}(w) = O((1-|w|)^{(\alpha+2)q/p}), \ |w| \to 1.$$

The analogous statements hold for compactness if the 'big-oh' condition is replaced by 'little-oh'.

If $0 < \sigma < \gamma$ and $w \in D$, $w \neq \Phi(0)$, then $N_\gamma(w) \leq (N_\sigma(w))^{\gamma/\sigma}$. If Φ is of finite valence, then there is a constant C such that $N_\sigma(w) \leq C(N_\gamma(w))^{\sigma/\gamma}$. Thus $(N_\gamma(w))^\sigma \approx (N_\sigma(w))^\gamma$ for such functions. See [**14**] for a discussion of these inequalities.

COROLLARY 4.3. *Let $p \geq 1, \alpha \geq -1$ and $\beta > 0$. Suppose that $C_\Phi : A^p_\alpha \to L^{p\beta}(m)$ is order bounded and $(\alpha+2)\beta \geq 1$. Then $C_\Phi : A^p_\gamma \to A^{p(\alpha+2)\beta}_\gamma$ is compact for any $\gamma \geq -1$.*

PROOF. Since $C_\Phi : A^p_\alpha \to L^{p\beta}(m)$ is order bounded, Domenig's result yields $1/(1-\mid \Phi^* \mid) \in L^{(\alpha+2)\beta}(m)$. By Theorem 1.2, $C_\Phi : H^p \to L^{p(\alpha+2)\beta}(m)$ is order bounded. Since $(\alpha+2)\beta \geq 1$, Theorem 4.1 (Part 1) yields that $C_\Phi : H^p \to H^{p(\alpha+2)\beta}$ is compact. This completes the proof in the case $\gamma = -1$.

Let $\gamma > -1$. Riedl's characterization yields

$$N_1(w) = o(\,(1-\mid w \mid)^{(\alpha+2)\beta}\,)$$

as $\mid w \mid\to 1$. Since $\gamma + 2 > 1$, the remarks before the corollary yield

$$N_{\gamma+2}(w) = o(\,(1-\mid w \mid)^{(\alpha+2)\beta(\gamma+2)}\,)$$

as $\mid w \mid\to 1$. By Smith's characterization, $C_\Phi : A^p_\gamma \to A^{p(\alpha+2)\beta}_\gamma$ is compact. □

COROLLARY 4.4. *Let $\alpha \geq -1$ and $\beta > 0$. Suppose that Φ is of finite valence and $C_\Phi : A^p_\alpha \to A^{p\gamma}_\alpha$ is bounded for some $p \geq 1$ and $\gamma > \beta + 1$. Then $C_\Phi : A^p_\alpha \to L^{p\beta/(\alpha+2)}(m)$ is order bounded for all $p \geq 1$.*

PROOF. Because of Theorem 4.1 (Part 2), we may assume $\alpha > -1$. Since $C_\Phi : A^p_\alpha \to A^{p\gamma}_\alpha$ is bounded,

$$N_{\alpha+2}(w) = O(\,(1-\mid w \mid)^{(\alpha+2)\gamma}\,) \text{ as } \mid w \mid\to 1.$$

The valence hypothesis now yields $N_1(w) = O(\,(1-\mid w \mid)^\gamma\,)$ as $\mid w \mid\to 1$ and thus by Riedl's result $C_\Phi : H^p \to H^{p\gamma}$ is bounded. Since $\beta < \gamma - 1$, Theorem 4.1 (Part 2) implies that $C_\Phi : H^p \to L^{\beta p}(m)$ is order bounded, and thus

$$1/(1-\mid \Phi^* \mid)^\beta \in L^1(m).$$

By Domenig's Theorem [**1**] this is equivalent to order boundedness of $C_\Phi : A^p_\alpha \to L^{\beta p/(\alpha+2)}(m)$. □

In the remainder of this work we assume that Ψ is analytic in D and Φ is an analytic self-map of D. The closing results characterize the weighted composition operators $W_{\Psi,\Phi} : A^p_\alpha \to A^q_\beta$ which are bounded or compact. Related results for C_Φ were given by W. Smith [**14**] in the case $0 < p \leq q$, and by Smith and L. Yang [**15**] in the case $0 < q < p$. Let $\alpha > -1$ and let $dA_\alpha(z)$ denote the measure $(1-\mid z \mid^2)^\alpha\, dA(z)$. Smith and Yang showed that if $q < p$ and $\alpha > -1$, then

$$C_\Phi : A^p_\alpha \to A^q_\beta \text{ is bounded } \Leftrightarrow C_\Phi : A^p_\alpha \to A^q_\beta \text{ is compact}$$

$$\frac{N_{\beta+2}(z)}{(1-\mid z \mid^2)^{\alpha+2}} \in L^{p/(p-q)}(A_\alpha).$$

Let $a \in D$. In the rest of this section, $D(a)$ denotes the pseudohyberbolic disc centered at a with radius $1/8$, that is,

$$D(a) = \{z : \left| \frac{a-z}{1-\bar{a}z} \right| < 1/8\}.$$

The following lemma is well known.

LEMMA 4.5. *(1)* $| 1 - \bar{a}w | \approx 1 - | a |^2$ *for* $w \in D(a)$.
(2) $1 - | w |^2 \approx 1 - | a |^2$ *for* $w \in D(a)$.

THEOREM 4.6. *Let* $1 \leq p \leq q$ *and let* $\alpha, \beta > -1$. *Assume that* $\Psi \in A^q_\beta$ *and let* Φ *be an analytic self-map of the disc. The following are equivalent.*
(1) $W_{\Psi,\Phi} : A^p_\alpha \to A^q_\beta$ *is bounded.*
(2) $\int_{\Phi^{-1}(D(a))} | \Psi |^q \, dA_\beta = O((1 - | a |^2)^{(\alpha+2)q/p})$ *as* $| a | \to 1$.
Compactness is characterized by the analogous little-oh condition.

PROOF. First assume that $W_{\Psi,\Phi} : A^p_\alpha \to A^q_\beta$ is bounded. For $a \in D$, let

$$f_a(z) = \frac{(1 - | a |^2)^{(\alpha+2)/p}}{(1 - \bar{a}z)^{2(\alpha+2)/p}}, | z | < 1.$$

Since $\| f_a \|_{A^p_\alpha} \approx 1$ [**14**], there is a constant C with

$$C \geq (1 - | a |^2)^{(\alpha+2)q/p} \int_{\Phi^{-1}(D(a))} \frac{| \Psi(z) |^q}{| 1 - \bar{a}\Phi(z) |^{2(\alpha+2)q/p}} \, dA_\beta(z)$$

for all $a \in D$. The first estimate in Lemma 4.5 now yields the result.

Assume that the second condition holds, that is, there exists r_0, $0 < r_0 < 1$, and a constant C such that

$$\int_{\Phi^{-1}(D(w))} | \Psi(z) |^q \, dA_\beta < C \, (1 - | w |^2)^{(\alpha+2)q/p} \text{ for } | w | > r_0.$$

By the closed graph theorem, it is enough to show that $\| W_{\Psi,\Phi}(f) \|_{A^q_\beta} < \infty$ for all $f \in A^p_\alpha$. A standard estimate yields a positive constant C such that

$$\| W_{\Psi,\Phi}(f) \|^q_{A^q_\beta} \leq \int_D | \Psi(z) |^q \frac{C}{(1 - | \Phi(z) |^2)^2} \int_{D(\Phi(z))} | f(w) |^q \, dA(w) \, dA_\beta(z).$$

Since $w \in D(\Phi(z)) \Leftrightarrow z \in \Phi^{-1}(D(w))$, Fubini's theorem and Lemma 4.5 (Part 2) yield

$$\| W_{\Psi,\Phi}(f) \|^q_{A^q_\beta} \leq C \int_D | f(w) |^q \frac{\int_{\Phi^{-1}(D(w))} | \Psi(z) |^q \, dA_\beta(z)}{(1 - | w |^2)^2} \, dA(w).$$

By Lemma 1.3, $| f(w) | \leq C \| f \|_{A^p_\alpha} (1 - | w |)^{-(\alpha+2)/p}$ and it follows that

$$\| W_{\Psi,\Phi}(f) \|^q_{A^q_\beta} \leq C \, \| f \|^{q-p}_{A^p_\alpha} \int_D | f(w) |^p \frac{\int_{\Phi^{-1}(D(w))} | \Psi(z) |^q \, dA_\beta(z)}{(1 - | w |^2)^{(\alpha+2)q/p}} \, dA_\alpha(w).$$

Since $\Psi \in A^q_\beta$ and $f \in A^p_\alpha$,

$$\int_{|w| \leq r_0} | f(w) |^p \frac{\int_{\Phi^{-1}(D(w))} | \Psi(z) |^q \, dA_\beta(z)}{(1 - | w |^2)^{(\alpha+2)q/p}} \, dA_\alpha(w)$$

$$\leq \frac{1}{(1 - r_0^2)^{(\alpha+2)q/p}} \int_D | f(w) |^p \int_D | \Psi(z) |^q \, dA_\beta(z) \, dA(w) < \infty.$$

The hypothesis implies that

$$\int_{r_0<|w|<1} | f(w) |^p \frac{\int_{\Phi^{-1}(D(w))} | \Psi(z) |^q \, dA_\beta(z)}{(1- | w |^2)^{(\alpha+2)q/p}} \, dA_\alpha(w) \leq C \| f \|^p_{A^p_\alpha} .$$

Thus $\| W_{\Psi,\Phi}(f) \|_{A^q_\beta} < \infty$, and $W_{\Psi,\Phi} : A^p_\alpha \to A^q_\beta$ is bounded.

The proof of the statement about compactness involves similar estimates and the standard criterion for compactness. The details are omitted. □

In the case $q < p$, the characterization will make use of work of D. Luecking. In [**8**], Luecking used Khinchine's inequality and other techniques to prove the following result on the Bergman space $A^p = A^p_0$. The result is stated here in the more general setting of the spaces A^p_α. See [**8**] for the ideas involved in the proof for $\alpha \neq 0$. The restriction $\alpha > -1$ is needed for the construction of suitable test functions to be used in the proof of the last theorem.

THEOREM 4.7 (Luecking). *Let $0 < q < p$ and let $\alpha > -1$. Let μ be a positive measure on the disc, and let $L(z) = \mu(D(z)) \, (1- | z |^2)^{-(\alpha+2)}$. The following are equivalent.*

(1) *There is a constant C such that*

$$(\int_D | f |^q \, d\mu)^{1/q} \leq C \| f \|_{A^p_\alpha} \textit{ for all } f \in A^p_\alpha.$$

(2) $L \in L^{p/(p-q)}(A_\alpha)$.

THEOREM 4.8. *Let $1 \leq q < p$ and let $\alpha > -1$. Let $\Psi \in A^q_\beta$. For $| z |< 1$, define*

$$L(z) = \frac{\int_{\Phi^{-1}(D(z))} | \Psi(w) |^q \, dA_\beta(w)}{(1- | z |^2)^{\alpha+2}}.$$

The following are equivalent.

(1) $W_{\Psi,\Phi} : A^p_\alpha \to A^q_\beta$ *is bounded.*

(2) $W_{\Psi,\Phi} : A^p_\alpha \to A^q_\beta$ *is compact.*

(3) $L \in L^{p/(p-q)}(A_\alpha)$.

PROOF. First suppose that $L \in L^{p/(p-q)}(A_\alpha)$, $\| f_n \|_{A^p_\alpha} \leq C$ and $f_n \to 0$ uniformly on compact subsets of the disc. To prove (2) it will be enough to show that $\| W_{\Psi,\Phi}(f_n) \|_{A^q_\beta} \to 0$ as $n \to \infty$. An argument as in the proof of Theorem 4.6 yields

$$\| W_{\Psi,\Phi}(f_n) \|^q_{A^q_\beta} \leq C \int_D | f_n(w) |^q \, L(w) \, dA_\alpha(w).$$

Let $\epsilon > 0$ be given. The hypothesis (3) implies that there exists r, $0 < r < 1$, such that

$$\int_{r<|z|<1} L(z)^{p/(p-q)} \, dA_\alpha < \epsilon^{p/(p-q)}.$$

Since $\| f_n \|_{A^p_\alpha} \leq C$ for all n, Hölder's inequality yields

$$\int_{r<|z|<1} | f_n(w) |^q \, L(w) \, dA_\alpha(w)$$

$$\leq \| f_n \|^q_{A^p_\alpha} \, (\int_{r<|z|<1} L(w)^{p/(p-q)} \, dA_\alpha)^{(p-q)/p} \leq C^q \epsilon \text{ for all } n.$$

Since $f_n \to 0$ uniformly on $\{z : | z | \leq r\}$ and since $\Psi \in A^q_\beta$,

$$\int_{|z|\leq r} | f_n(w) |^q \; L(w) \, dA_\alpha \; \leq \; \epsilon \, (1-r^2)^{-(\alpha+2)} \; \| \Psi \|^q_{A^q_\beta} \; \int_D dA_\alpha(z)$$

for all large n. Thus $\| W_{\Psi,\Phi}(f_n) \|_{A^q_\beta} \to 0$ as $n \to \infty$. The proof of (3) $\Rightarrow$ (2) is complete.

It is clear that (2) $\Rightarrow$ (1).

Finally assume that $W_{\Psi,\Phi} : A^p_\alpha \to A^q_\beta$ is bounded. Thus there is a constant C such that

$$\left(\int_D | \Psi(z) |^q \, | (f \circ \Phi)(z) |^q \; dA_\beta(z) \right)^{1/q} \; \leq \; C \; \| f \|_{A^p_\alpha} .$$

Let ν be the measure defined by $d\nu(z) \; = | \Psi(z) |^q \; dA_\beta(z)$. Then ν is a positive measure on D and the previous expression can be rewritten as

$$\left(\int_D | f(w) |^q \; d(\nu\Phi^{-1}) \right)^{1/q} \; \leq \; C \; \| f \|_{A^p_\alpha} .$$

Luecking's result finishes the proof. □

The restriction $\alpha > -1$ in Theorem 4.8 can not be removed. To see this, let $\Psi = 1, \Phi(z) = z$, and consider the operator $W_{\Psi,\Phi} : H^2 \to H^1$, which is bounded but not compact.

References

[1] T. Domenig, *Order Bounded and p-summing Composition Operators*, Contemporary Mathematics, vol. 213 (1998), pp. 27-41.

[2] P. L. Duren, *Theory of H^p Spaces*, Academic Press, New York, 1970.

[3] P. L. Duren and A. Schuster, *Bergman Spaces*, Amer. Math. Soc., 2004.

[4] H. Hedenmalm, B. Korenblum and K. Zhu, *Theory of Bergman Spaces*, Springer, New York, 2000.

[5] H. Hunziker, *Kompositionsoperatoren auf klassichen Hardyraumen*, Dissertation, Universitat Zurich, 1989.

[6] H. Hunziker and H. Jarchow, *Composition operators which improve integrability*, Math. Nachr. 152 (1991), pp. 83-99.

[7] H. Jarchow and R. Riedl, *Factorization of composition operators through Bloch type spaces*, Illinois Journal of Math. 39 (1995), pp. 431-440.

[8] D. Luecking, *Embedding theorems for spaces of analytic functions via Khinchine's inequality*, Mich. Math. J. 40 (1993), pp. 333-358.

[9] K. Madigan, *Composition operators on analytic Lipschitz spaces*, Proc. Amer. Math. Soc. 119 (1993), pp. 465-473.

[10] A. Montes-Rodriguez, *The essential norm of a composition operator on Bloch spaces*, Pacific Journal of Mathematics 188 (1999), pp. 339-351.

[11] A. Montes-Rodriguez, *Weighted composition operators on weighted Banach spaces*, J. London Math. Soc. (2) 61 (2000), pp. 872-884.

[12] R. Riedl, *Composition operators and geometric properties of analytic functions*, Dissertation, Universitat Zurich, 1994.

[13] J. H. Shapiro and P. Taylor, *Compact, nuclear and Hilbert-Schmidt composition operators on H^2*, Indiana Univ. Math. J. 23 (1973), pp. 471-496.

[14] W. Smith, *Composition operators between Bergman and Hardy spaces*, Trans. Amer. Math. Soc. 348 (1996), pp. 2331-2348.

[15] W. Smith and L. Yang, *Composition Operators that improve integrability on weighted Bergman Spaces*, Proc. Amer. Math. Soc. 126 (1998), pp. 411-420.

[16] K. Zhu, *Operator Theory in Function Spaces*, Marcel Dekker, Inc, New York, 1990.

Department of Mathematics and Statistics, University of New Hampshire, Durham, New Hampshire, 03824

E-mail address: rah2@unh.edu

Contemporary Mathematics
Volume **454**, 2008

Fractional Cauchy Transforms and Composition

T.H. MacGregor

1. Introduction

Let $\mathcal{D} = \{z \in \mathbb{C} : | z | < 1\}$ and let $T = \{z \in \mathbb{C} : | z | = 1\}$. Let $\mathcal{M}$ denote the set of complex-valued Borel measures on T. For each positive real number α, let $\mathcal{F}_\alpha$ denote the set of functions f such that there exists $\mu \in \mathcal{M}$ for which

$$f(z) = \int_T \frac{1}{(1 - \bar{\zeta} z)^\alpha} \, d\mu(\zeta) \tag{1}$$

for $| z | < 1$. The power function in (1) is the principal branch.

If $f \in \mathcal{F}_\alpha$, let

$$\|f\|_{\mathcal{F}_\alpha} = \inf_{\mu \in \mathcal{M}} \|\mu\| \tag{2}$$

where $\|\mu\|$ denotes the total variation of μ and μ in (2) varies over all members of $\mathcal{M}$ for which (1) holds. This defines a norm on $\mathcal{F}_\alpha$ and $\mathcal{F}_\alpha$ is a Banach space with respect to this norm.

A function given by (1) is called a fractional Cauchy transform of order α. When $\alpha = 1$ this gives the Cauchy transform of a measure supported on T. The book [10] is an introduction to the research on the families $\mathcal{F}_\alpha$.

For $\alpha > 0$ let $\mathcal{C}_\alpha$ denote the set of functions $\varphi : \mathcal{D} \to \mathcal{D}$ such that the composition $f \circ \varphi \in \mathcal{F}_\alpha$ for every $f \in \mathcal{F}_\alpha$. If $\varphi \in \mathcal{C}_\alpha$ then the mapping $f \longmapsto f \circ \varphi$ is a continuous linear operator on $\mathcal{F}_\alpha$. Let $\|\varphi\|_{\mathcal{C}_\alpha}$ denote the norm of this operator. The identity function belongs to $\mathcal{F}_\alpha$ for every $\alpha > 0$ and hence $\mathcal{C}_\alpha \subset \mathcal{F}_\alpha$. In particular, each member of $\mathcal{C}_\alpha$ is analytic in $\mathcal{D}$.

This paper concerns the problem of describing the functions which belong to $\mathcal{C}_\alpha$. A survey of known facts about this problem is given and some new results are presented.

In [4] Cima and Matheson study composition operators on the family of Cauchy transforms where the emphasis is on questions relating to the compactness and the weak compactness of these operators.

2000 Mathematics Subject Classification. 30E20.

2. Basic facts about $\mathcal{C}_\alpha$

Theorem 1. If $\varphi : \mathcal{D} \to \mathcal{D}$ is analytic, then $\varphi \in \mathcal{C}_\alpha$ for all $\alpha \geq 1$.

Theorem 1 was proved by Bourdon and Cima in [2] when $\alpha = 1$ and by Hibschweiler and MacGregor in [8] when $\alpha > 1$. The arguments rely on the classical result of Herglotz and Riesz about functions having a positive real part and on a generalization concerning the family of functions subordinate to $F(z) = \dfrac{1}{(1-z)^\alpha}$.

Theorem 2. If $\varphi \in \mathcal{C}_\alpha$ and $\beta > \alpha$ then $\varphi \in \mathcal{C}_\beta$.

Theorem 2 was proved by Hibschweiler in [7]. A critical step in the argument uses Theorem 1 with $\alpha > 1$.

Theorem 3. If φ is an analytic function that maps $\mathcal{D}$ one-to-one onto $\mathcal{D}$, then $\varphi \in \mathcal{C}_\alpha$ for all $\alpha > 0$.

Theorem 3 was proved by Hibschweiler and MacGregor in [8]. This property of conformal automorphisms of $\mathcal{D}$ serves as a lemma for various arguments.

3. Necessary conditions

As mentioned earlier, $\mathcal{C}_\alpha \subset \mathcal{F}_\alpha$. Since the Taylor coefficients of a function in $\mathcal{F}_\alpha$ satisfy

$$a_n = O(n^{\alpha-1}) \quad (n \to \infty) \tag{3}$$

this yields examples of analytic functions $\varphi : \mathcal{D} \to \mathcal{D}$ which do not belong to $\mathcal{C}_\alpha$. If $0 < \alpha < 1$ we may let $\varphi(z) = \sum_{n=1}^{\infty} b_n z^{n^p}$ where p is a positive integer depending on α and $\{b_n\}$ is a suitable sequence with $\sum_{n=1}^{\infty} | b_n | \leq 1$. Likewise (3) implies that $\varphi \notin \mathcal{C}_\alpha$ for all $\alpha (0 < \alpha < 1)$ when $\varphi(z) = \epsilon \sum_{n=1}^{\infty} \frac{1}{n^2} z^{2^n} (| z | < 1)$ and ϵ is sufficiently small and $\epsilon \neq 0$.

The Taylor coefficients of members of $\mathcal{C}_\alpha$ also satisfy the following condition.

Theorem 4. If $0 < \alpha < 1, \varphi \in \mathcal{C}_\alpha$ and $\varphi(z) = \sum_{n=0}^{\infty} a_n z^n (| z | < 1)$, then

$$\sum_{n=0}^{\infty} (n+1)^{1-\alpha} | a_n |^2 < \infty. \tag{4}$$

Theorem 4 is a consequence of the more general result in [6; see Theorem 1, p. 163] that (4) holds if $0 < \alpha < 1$ and $\varphi \in H^\infty \cap \mathcal{F}_\alpha$.

4. The case $\|\varphi\|_{H^\infty} < 1$

For $\alpha > 0$ let $\mathcal{B}_\alpha$ denote the set of functions f that are analytic in $\mathcal{D}$ and satisfy

$$\int_0^1 \int_{-\pi}^{\pi} | f'(re^{i\theta}) | (1-r)^{\alpha-1} d\theta \, dr < \infty. \tag{5}$$

In [6; see Lemma 2, p. 160] it was shown that $\mathcal{B}_\alpha \subset \mathcal{F}_\alpha$. This gives an analytic way of showing membership in $\mathcal{F}_\alpha$.

Theorem 5. If $0 < \alpha < 1, \varphi \in \mathcal{B}_\alpha$ and $\sup_{|z|<1} | \varphi(z) | < 1$ then $\varphi \in \mathcal{C}_\alpha$.

Theorem 5 was proved by Hibschweiler in [7; see Corollary 2.1, p. 62].

Below we present two results about membership in $\mathcal{B}_\alpha$. Combined with Theorem 5 this yields sufficient conditions for membership in $\mathcal{C}_\alpha$ when $\sup_{|z|<1} | \varphi(z) |< 1$.

Theorem 6. Suppose that $f \in H^1$ and let $F(\theta) = \lim_{r\to 1-} f(re^{i\theta})$ and $D(\theta,\lambda) = F(\theta+\lambda) - 2F(\theta) + F(\theta-\lambda)$ for suitable values of θ and λ in $[-\pi,\pi]$. If $0<\alpha<1$ and

$$\int_{-\pi}^{\pi}\int_0^{\pi} \frac{| D(\theta,\lambda) |}{\lambda^{2-\alpha}} d\lambda\, d\theta < \infty \tag{6}$$

then $f \in \mathcal{B}_\alpha$.

A proof of Theorem 6 follows from the result shown in [10; see p. 139-141]: If $f \in H^1$ and $0<\alpha<2$ then there is a constant A such that

$$\int_0^1 | f'(re^{i\theta}) | (1-r)^{\alpha-1} dr \le A \int_0^{\pi} \frac{| D(\theta,\lambda) |}{\lambda^{2-\alpha}} d\lambda. \tag{7}$$

We see from (7) and Tonelli's theorem that (6) implies (5).

Theorem 7. Let $0<\alpha<1$ and suppose that $\varphi(z) = \sum_{n=0}^{\infty} a_n z^n (| z |< 1)$ and

$$\sum_{n=0}^{\infty} (n+1)^{1-\alpha} | a_n |< \infty. \tag{8}$$

Then $\varphi \in \mathcal{B}_\beta$ for $\beta > \alpha$.

Proof of Theorem 7: If $\gamma>0$ then there is a positive constant A such that

$$k^\gamma r^k \le \frac{A}{(1-r)^\gamma} \tag{9}$$

for $0<r<1$ and $k=1,2,\ldots$. This implies that if $0<\alpha<1, | z |= r$ and $0<r<1$ then

$$| \varphi'(z) | \le \sum_{n=1}^{\infty} n \mid a_n \mid r^{n-1} \le | a_1 | + 2^\alpha \sum_{n=2}^{\infty} (n-1)^\alpha r^{n-1} n^{1-\alpha} \mid a_n \mid \le | a_1 | + \frac{2^\alpha A}{(1-r)^\alpha} \sum_{n=2}^{\infty} n^{1-\alpha} | a_n |.$$

Hence the assumption (8) implies that there is a positive constant B such that $| \varphi'(z) | \le \dfrac{B}{(1-r)^\alpha}$. If $\beta>\alpha$ this yields

$$\int_0^1 \int_{-\pi}^{\pi} | \varphi'(re^{i\theta}) | (1-r)^{\beta-1} d\theta dr \le B \int_0^1 (1-r)^{\beta-\alpha-1} dr < \infty.$$

Therefore $\varphi \in \mathcal{B}_\beta$ for $\beta>\alpha$. □

The following theorem gives a sufficient condition for membership in $\mathcal{C}_\alpha$ for all $\alpha>0$ and only depends on the Taylor coefficients.

Theorem 8. Suppose that the function $\varphi : \mathcal{D} \to \mathcal{D}$ is given by $\varphi(z) = \sum_{n=0}^{\infty} a_n z^n (| z |< 1)$ and

$$\sum_{n=0}^{\infty} (n+1) | a_n |< \infty. \tag{10}$$

Then $\varphi \in \mathcal{C}_\alpha$ for all $\alpha>0$.

Theorem 8 was proved in [10; see p. 200]. The argument relies on Theorem 1 with $\alpha > 1$ and on a result about the multipliers of $\mathcal{F}_\alpha$ involving Taylor coefficients. It is not known whether (8) implies $\varphi \in \mathcal{C}_\alpha$ where $0 < \alpha < 1$.

Our knowledge of which univalent functions belong to $\mathcal{C}_\alpha$ is quite limited. The main fact is stated below. It was proved in [10; see Theorem 9.10, p. 214] using Theorem 5.

Theorem 9. Let $\alpha_0 = \frac{1}{2} - \frac{1}{320}$. If the function φ is analytic and univalent in $\mathcal{D}$ and $\sup\limits_{|z|<1} | \varphi(z) | < 1$ then $\varphi \in \mathcal{C}_\alpha$ for all $\alpha > \alpha_0$.

The next two theorems concern constant multiples of a function and membership in $\mathcal{C}_\alpha$.

Theorem 10. Suppose that $\alpha > 0, \varphi \in \mathcal{C}_\alpha$ and $\psi = b\varphi$ where $| b | \leq 1$. Then $\psi \in \mathcal{C}_\alpha$.

Proof of Theorem 10: Suppose that $\alpha > 0$ and $| b | \leq 1$. For $| z | < 1$ let $F(z) = \frac{1}{(1-z)^\alpha}$, and let $| \zeta | = 1$. Since F is analytic in $\mathcal{D}$ there is a probability measure $\mu \in \mathcal{M}$ such that

$$F(b\bar{\zeta}z) = \int_T F(\bar{\sigma}z) d\mu(\sigma)$$

for $| z | < 1$ [10; see p. 21]. The equation

$$\frac{1}{(1-\bar{\zeta}bz)^\alpha} = \int_T \frac{1}{(1-\bar{\sigma}z)^\alpha} d\mu(\sigma) (| z | < 1)$$

and the fact that μ is a probability measure imply that $\| \frac{1}{(1-\bar{\zeta}bz)^\alpha} \|_{\mathcal{F}_\alpha} = 1$.

Suppose that $\varphi \in \mathcal{C}_\alpha$ and let $M = \|\varphi\|_{\mathcal{C}_\alpha}$. Then $F(b\bar{\zeta}\varphi(z)) \in \mathcal{F}_\alpha$ and $\|F(b\bar{\zeta}\varphi(z))\|_{\mathcal{F}_\alpha} \leq M\|F(b\bar{\zeta}z)\|_{\mathcal{F}_\alpha} = M$. Let $\psi = b\varphi$. We have $\frac{1}{(1-\bar{\zeta}\psi)^\alpha} \in \mathcal{F}_\alpha$ and $\| \frac{1}{(1-\bar{\zeta}\psi)^\alpha} \|_{\mathcal{F}_\alpha} \leq M$. The last inequality holds for all $\zeta (| \zeta | = 1)$. This implies that $f \circ \psi \in \mathcal{F}_\alpha$ for all $f \in \mathcal{F}_\alpha$ [7; see p. 59]. Therefore $\psi \in \mathcal{C}_\alpha$. □

Let $\mathcal{M}_\alpha$ denote the set of functions f such that $fg \in \mathcal{F}_\alpha$ for every $g \in \mathcal{F}_\alpha$. If $f \in \mathcal{M}_\alpha$ then the mapping $g \mapsto fg$ is a continuous linear operator on $\mathcal{F}_\alpha$. We let $\|f\|_{\mathcal{M}_\alpha}$ denote the norm of this operator. Since the constant function 1 belongs to $\mathcal{F}_\alpha$ for every $\alpha > 0$, we obtain $\mathcal{M}_\alpha \subset \mathcal{F}_\alpha$. The family of multipliers $\mathcal{M}_\alpha$ has been extensively studied [10; see Chapters 6 and 7]. Members of $\mathcal{M}_\alpha$ have a number of properties including being bounded.

Theorem 11. Suppose that $\alpha > 0, f \in \mathcal{M}_\alpha, f \neq 0$ and b is any complex number such that $| b | < \frac{1}{\|f\|_{\mathcal{M}_\alpha}}$. Then $bf \in \mathcal{C}_\alpha$.

Proof of Theorem 11: Let $M = \|f\|_{\mathcal{M}_\alpha}$. The assumption $f \neq 0$ implies that $M > 0$. We have $\|fg\|_{\mathcal{F}_\alpha} \leq M\|g\|_{\mathcal{F}_\alpha}$ for all $g \in \mathcal{F}_\alpha$.

The constant function 1 belongs to $\mathcal{F}_\alpha$ and $\|1\|_{\mathcal{F}_\alpha} = 1$. Hence $f \in \mathcal{F}_\alpha$ and $\|f\|_{\mathcal{F}_\alpha} \leq M$. Also $f^2 \in \mathcal{F}_\alpha$ and $\|f^2\|_{\mathcal{F}_\alpha} \leq M\|f\|_{\mathcal{F}_\alpha} \leq M^2$, and, in general, $f^k \in \mathcal{F}_\alpha$ and $\|f^k\|_{\mathcal{F}_\alpha} \leq M^k$ for $k = 1, 2, \ldots$.

Suppose that $| b |< \frac{1}{M}$. For $k = 1, 2, \ldots$ let $A_k(\alpha)$ denote the binomial coefficients defined by

$$\frac{1}{(1-z)^\alpha} = \sum_{k=0}^{\infty} A_k(\alpha) z^k (| z |< 1).$$

Suppose that $| \zeta |= 1$ and for $n = 1, 2, \ldots$ let

$$p_n(z) = \sum_{k=0}^{n} A_k(\alpha)[b\bar{\zeta} f(z)]^k (| z |< 1).$$

Then $p_n \in \mathcal{F}_\alpha$ and

$$\begin{aligned} \|p_n\|_{\mathcal{F}_\alpha} &\leq \sum_{k=0}^{n} A_k(\alpha) \, | b |^k \, \|f^k\|_{\mathcal{F}_\alpha} \\ &\leq \sum_{k=0}^{n} A_k(\alpha) \, | b |^k \, M^k \\ &\leq \sum_{k=0}^{\infty} A_k(\alpha) \, | b |^k \, M^k = \frac{1}{(1- | b | M)^\alpha} \equiv P \end{aligned}$$

Since $| b |< \frac{1}{M}, P < \infty$.

From $f \in \mathcal{M}_\alpha$ it follows that $f \in H^\infty$ and $\|f\|_{H^\infty} \leq M$ [9; see p. 380]. (The last inequality and Theorem 1 give another proof of this theorem when $\alpha \geq 1$.) Hence

$$| b\bar{\zeta} f(z) | \leq | b | M < 1 \quad \text{for} \quad | z |< 1.$$

Therefore $p_n(z) \to \sum_{k=0}^{\infty} A_k(\alpha)[b\bar{\zeta} f(z)]^k$ uniformly in $\mathcal{D}$ as $n \to \infty$. Since $\|p_n\|_{\mathcal{F}_\alpha} \leq P$ for $n = 1, 2, \ldots$ and $p_n(z) \to \sum_{k=0}^{\infty} A_k(\alpha)[b\bar{\zeta} f(z)]^k$ for each z in $\mathcal{D}$, it follows that $\sum_{k=0}^{\infty} A_k(\alpha)[b\bar{\zeta} f(z)]^k$ belongs to $\mathcal{F}_\alpha$ and $\|\sum_{k=0}^{\infty} A_k(\alpha)[b\bar{\zeta} f(z)]^k\|_{\mathcal{F}_\alpha} \leq P$ [10; see Lemma 7.9, p. 146]. We have shown that $\frac{1}{[1-\bar{\zeta} b f(z)]^\alpha} \in \mathcal{F}_\alpha$ and $\|\frac{1}{[1-\bar{\zeta} b f(z)]^\alpha}\|_{\mathcal{F}_\alpha} \leq P$ for all $\zeta (| \zeta |= 1)$. This implies that $bf \in \mathcal{C}_\alpha$. □

5. Angular derivatives

Our study of $\mathcal{C}_\alpha$ relates to results of Julia about the angular derivative of a bounded analytic function.

Theorem 12. Suppose that the function $\varphi : \mathcal{D} \to \mathcal{D}$ is analytic. Let $\sigma \in T$. Then for each $w \in T$ the non-tangential limit

$$\beta(w) \equiv \lim_{z \to w} \frac{\varphi(z) - \sigma}{z - w} \tag{11}$$

exists or equals ∞. Let $\Lambda = \{w \in T : \beta(w) \neq \infty\}$ and let $\gamma(w) = \frac{w}{\sigma}\beta(w)$. If $w \in \Lambda$ than $\gamma(w)$ is a positive real number and the non-tangential limit $\lim_{z \to w} \varphi'(z)$

exists and equals $\beta(w)$. If Λ is non-vacuous then either Λ is finite or Λ is countably infinite, and

$$\sum_{w\in\Lambda}\frac{1}{\gamma(w)} \le \frac{1+\mid\varphi(0)\mid}{1-\mid\varphi(0)\mid}. \tag{12}$$

Except for the last sentence in Theorem 12, this result and related facts due to Julia can be found in [1; see p. 7], [3; see p. 23] and [5; see p. 43]. We present an argument which also yields the last assertion.

Proof of Theorem 12: Suppose that the function $\varphi : \mathcal{D} \to \mathcal{D}$ is analytic and let $\mid \sigma \mid = 1$. For $\mid z \mid < 1$ let

$$p(z) = \frac{1}{1-\bar{\sigma}\varphi(z)} \tag{13}$$

Set $b = \mathrm{Re}p(0)$ and $c = \mathrm{Im}p(0)$. The function p is analytic in $\mathcal{D}$ and $\mathrm{Re}p(z) > 1/2$ for $\mid z \mid < 1$. The function

$$q = \frac{b-1}{2b-1} + \frac{p-ic}{2b-1} \tag{14}$$

is analytic in $\mathcal{D}$, $\mathrm{Re}q(z) > 1/2$ for $\mid z \mid < 1$ and $q(0) = 1$. The Herglotz-Riesz formula yields

$$q(z) = \int_T \frac{1}{1-\bar{\zeta}z}d\mu(\zeta) \quad (\mid z \mid < 1) \tag{15}$$

where μ is a probability measure on T. Let $e = 2b-1$ and $f = 1-b+ic$. Then $e > 0$ and (15) and (14) yield

$$p(z) = e\int_T \frac{1}{1-\bar{\zeta}z}d\mu(\zeta) + f \quad (\mid z \mid < 1). \tag{16}$$

Suppose that $w \in T$. Then

$$(1-\bar{w}z)p(z) = e\int_T \frac{1-\bar{w}z}{1-\bar{\zeta}z}d\mu(\zeta) + f(1-\bar{w}z) \tag{17}$$

for $\mid z \mid < 1$. Let S denote a Stolz angle in $\mathcal{D}$ with vertex w. The integrand in (17) is bounded for $\zeta \in T$ and $z \in S$ and it equals 1 if $\zeta = w$ and it tends to zero as $z \to w$ if $\zeta \neq w$. The bounded convergence theorem yields

$$\lim_{z\to w}(1-\bar{w}z)p(z) = e\mu(\{w\}) \tag{18}$$

where $z \in S$. Since $e > 0$ and μ is a non-negative measure, this limit is a non-negative real number and it is zero if and only if $\mu(\{w\}) = 0$. Let

$$\Lambda = \{w \in T : \mu(\{w\}) \neq 0\}. \tag{19}$$

If Λ is non-vacuous then either Λ is finite or Λ is countably infinite.

From (13) we obtain $\dfrac{\varphi(z)-\sigma}{z-w} = \dfrac{\bar{w}\sigma}{(1-\bar{w}z)p(z)}$. Hence what was shown above about

$$\lim_{z\to w}(1-\bar{w}z)p(z)$$

implies that the non-tangential limit $\lim\limits_{z\to w}\dfrac{\varphi(z)-\sigma}{z-w}$ is either ∞ or $\dfrac{\bar{w}\sigma}{e\mu(\{w\})}$ depending on whether $\mu(\{w\}) = 0$ or $\mu(\{w\}) > 0$, respectively. Hence the set Λ defined by (19) has the properties stated in the theorem. We see that if $w \in \Lambda$ then $\gamma(w) = \dfrac{1}{e\mu(\{w\})} > 0$. Also

$$\sum_{w\in\Lambda} \frac{1}{\gamma(w)} = \sum_{w\in\Lambda} e\mu(\{w\}) \le e\mu(T) = e = \mathrm{Re}\left\{\frac{1+\bar{\sigma}\varphi(0)}{1-\bar{\sigma}\varphi(0)}\right\} \le \frac{1+\mid\varphi(0)\mid}{1-\mid\varphi(0)\mid}.$$

This proves (12).

Finally we prove the assertion about the non-tangential limit of φ'. From (13) we obtain $\varphi'(z) = \dfrac{\sigma p'(z)}{p^2(z)}$ and hence (16) yields

$$\varphi'(z) = \frac{\sigma e \displaystyle\int_T \frac{\bar{\zeta}}{(1-\bar{\zeta}z)^2} d\mu(\zeta)}{\left[e \displaystyle\int_T \frac{1}{1-\bar{\zeta}z} d\mu(\zeta) + f\right]^2}.$$

This can be written

$$\varphi'(z) = \frac{\sigma e \displaystyle\int_T \bar{\zeta}\left(\frac{1-\bar{w}z}{1-\bar{\zeta}z}\right)^2 d\mu(\zeta)}{\left[e \displaystyle\int_T \frac{1-\bar{w}z}{1-\bar{\zeta}z} d\mu(\zeta) + f(1-\bar{w}z)\right]^2}.$$

Let w belong to Λ and let S be a Stolz angle in $\mathcal{D}$ with vertex w. Then both integrands in the last expression are bounded on S. The bounded convergence theorem yields

$$\lim_{z\to w} \varphi'(z) = \frac{\sigma e \bar{w}\mu(\{w\})}{[e\mu(\{w\})]^2} = \beta(w)$$

where $z \in S$. $\square$

Theorem 13. Suppose that $0 < \alpha < 1$ and $\varphi \in \mathcal{C}_\alpha$. Let $\sigma \in T$ and let Λ and $\gamma(w)$ be defined as in Theorem 12 where $\mid w \mid= 1$. If Λ is non-vacuous then

$$\sum_{w\in\Lambda} \frac{1}{[\gamma(w)]^\alpha} \le \|\varphi\|_{\mathcal{C}_\alpha} \tag{20}$$

Proof of Theorem 13: For $\mid z \mid< 1$ let $f(z) = \dfrac{1}{[1-\bar{\sigma}\varphi(z)]^\alpha}$. The assumption $\varphi \in \mathcal{C}_\alpha$ implies there exists $\nu \in \mathcal{M}$ such that

$$f(z) = \int_T \frac{1}{(1-\bar{\zeta}z)^\alpha} d\nu(\zeta) \tag{21}$$

for $\mid z \mid< 1$. Let $w \in T$ and let $0 < r < 1$. Then $(1-r)^\alpha f(rw) = \displaystyle\int_T \left[\frac{1-r}{1-r\bar{\zeta}w}\right]^\alpha d\nu(\zeta)$ and the bounded convergence theorem yields

$$\lim_{r\to 1-} (1-r)^\alpha f(rw) = \nu(\{w\}) \tag{22}$$

Let the function p be defined by (13). Then $(1-r)^\alpha f(rw) = [(1-r)p(rw)]^\alpha$. Hence (18) yields

$$\lim_{r\to 1-} (1-r)^\alpha f(rw) = [e\mu(\{w\})]^\alpha \tag{23}$$

where e and μ have the same meaning as in the proof of Theorem 12. From (22) and (23) we obtain

$$\nu(\{w\}) = [e\mu(\{w\})]^{\alpha} \tag{24}$$

In particular, if $w \in \Lambda$ then $\gamma(w) = \dfrac{1}{e\mu(\{w\})}$ and thus $\gamma^{\alpha}(w) = \dfrac{1}{\nu(\{w\})}$. This gives $\displaystyle\sum_{w\in\Lambda} \frac{1}{[\gamma(w)]^{\alpha}} = \sum_{w\in\Lambda} \nu(\{w\})$. Hence

$$\sum_{w\in\Lambda} \frac{1}{[\gamma(w)]^{\alpha}} \leq \|\nu\|. \tag{25}$$

We have $\|g \circ \varphi\|_{\mathcal{F}_\alpha} \leq \|\varphi\|_{\mathcal{C}_\alpha} \|g\|_{\mathcal{F}_\alpha}$ for every $g \in \mathcal{F}_\alpha$. If $g(z) = \dfrac{1}{(1-\bar{\sigma}z)^{\alpha}}$ then $g \in \mathcal{F}_\alpha$ and $\|g\|_{\mathcal{F}_\alpha} = 1$. Hence

$$\|f\|_{\mathcal{F}_\alpha} \leq \|\varphi\|_{\mathcal{C}_\alpha}. \tag{26}$$

Inequality (25) holds for all $\nu \in \mathcal{M}$ which represent f in $\mathcal{F}_\alpha$. Thus (26) yields (20). □

Next we show that there is an analytic function $\varphi : \mathcal{D} \to \mathcal{D}$ for which the non-tangential limits of φ' have prescribed values on any given countable subset of T.

Theorem 14. Let $\sigma \in T$. Suppose that $\{w_k\} (k = 1, 2, \ldots)$ is a sequence of distinct points on T and $\{\gamma_k\} (k = 1, 2, \ldots)$ is a sequence of positive real numbers such that

$$\sum_{k=1}^{\infty} \frac{1}{\gamma_k} < \infty. \tag{27}$$

Then there is an analytic function $\varphi : \mathcal{D} \to \mathcal{D}$ such that for $k = 1, 2, \ldots$ the non-tangential limit

$$\lim_{z\to w_k} \varphi(z) \tag{28}$$

exists and equals σ, the non-tangential limit

$$\beta_k \equiv \lim_{z\to w_k} \frac{\varphi(z) - \sigma}{z - w_k} \tag{29}$$

exists, and $\dfrac{w_k \beta_k}{\sigma} = \gamma_k$. Also the non-tangential limit

$$\lim_{z\to w_k} \varphi'(z) \tag{30}$$

exists and equals β_k for $k = 1, 2, \ldots$.

Proof of Theorem 14: Let $\delta_k = \dfrac{1}{\gamma_k}$ for $k = 1, 2, \ldots$. By assumption $\delta_k > 0$ and the series $\displaystyle\sum_{k=1}^{\infty} \delta_k$ converges. Let $\delta = \displaystyle\sum_{k=1}^{\infty} \delta_k$. Let e be any real number such that $e \geq \delta$ and let $\epsilon_k = \dfrac{\delta_k}{e}$ for $k = 1, 2, \ldots$. Then $\epsilon_k > 0$ and $\displaystyle\sum_{k=1}^{\infty} \epsilon_k = \frac{\delta}{e} \leq 1$. Let μ be any probability measure on T such that

$$\mu(\{w_k\}) = \epsilon_k \tag{31}$$

for $k = 1, 2, \ldots$. For $|z| < 1$ let

$$q(z) = \int_T \frac{1}{1 - \bar{\zeta} z} d\mu(\zeta). \tag{32}$$

Then q is analytic in $\mathcal{D}$, $\text{Re}q(z) > 1/2$ for $|z| < 1$ and $q(0) = 1$. Let c be any real number and let $p = eq + \dfrac{1-e}{2} + ic$. Then p is analytic in $\mathcal{D}$, $\text{Re}p(z) > 1/2$ for $|z| < 1$ and $p(0) = b + ic$ where $b = \dfrac{1+e}{2}$. For $|z| < 1$ let

$$\varphi(z) = \sigma\left(1 - \frac{1}{p(z)}\right). \tag{33}$$

Then φ is analytic in $\mathcal{D}$ and $|\varphi(z)| < 1$ for $|z| < 1$.

Let $w \in T$. Then (33) implies

$$\frac{w}{\sigma}\frac{\varphi(z) - \sigma}{z - w} = \frac{1}{(1 - \bar{w}z)p(z)} = \frac{1}{(1 - \bar{w}z)\left[eq(z) + \frac{1-e}{2} + ic\right]}$$

If S is a Stolz angle in $\mathcal{D}$ with vertex w_k then the argument given in the proof of Theorem 12 and (31) yield

$$\lim_{z \to w_k} (1 - \bar{w}_k z) q(z) = \epsilon_k \text{ for } k = 1, 2, \ldots$$

where $z \in S$. Hence

$$\lim_{z \to w_k} \frac{w_k}{\sigma}\frac{\varphi(z) - \sigma}{z - w_k} = \frac{1}{e\epsilon_k} = \frac{1}{\delta_k} = \gamma_k$$

where $z \in S$. This proves the assertions about (29). This implies the claims about (28). The assertions about (30) follow by the argument given in the proof of Theorem 12. □

The argument for Theorem 14 shows that φ is obtained from the probability measure μ which is only subject to the condition (31). This provides a variety of functions φ when $e > \delta$. If $e = \delta$ then μ is the measure supported on $\bigcup_{k=1}^{\infty}\{w_k\}$ having mass ϵ_k at w_k. If, in addition, we let $c = 0$, this yields

$$\varphi(z) = \sigma\frac{s(z) - 1}{s(z) + 1} \quad \text{where} \quad s(z) = \sum_{k=1}^{\infty} \delta_k \frac{1 + w_k z}{1 - w_k z}.$$

By choosing the sequence $\{\gamma_k\}$ such that (27) holds and $\sum_{k=1}^{\infty} \dfrac{1}{\gamma_k^{\alpha}} = \infty$ for a given α where $0 < \alpha < 1$, we obtain further examples of analytic functions $\varphi : \mathcal{D} \to \mathcal{D}$ for which $\varphi \notin \mathcal{C}_\alpha$.

References

[1] L.V. Ahlfors, Conformal Invariants, McGraw-Hill, New York, 1973.

[2] P. Bourdon and J.A. Cima, On integrals of Cauchy-Stieltjes type, *Houston J. Math* 14 (1988), 465-474.

[3] C. Caratheodory, Theory of Functions of a Complex Variable, Vol. 2, Chelsea, New York, 1954.

[4] J.A. Cima and A. Matheson, Cauchy transforms and composition operators, *Illinois J. Math* 42 (1998), 58-69.

[5] J.B. Garnett, Bounded Analytic Functions, Academic Press, New York, 1981.

[6] D.J. Hallenbeck, T.H. MacGregor and K. Samotij, Fractional Cauchy transforms, inner functions and multipliers, *Proc. London Math. Soc.* (3) 72 (1996), 157-187.
[7] R.A. Hibschweiler, Composition operators on spaces of Cauchy transforms, *Contemporary Math* 213 (1998), 57-63.
[8] R.A. Hibschweiler and T.H. MacGregor, Closure properties of families of Cauchy-Stieltjes transforms, *Proc. Amer. Math. Soc.* 105 (1989), 615-621.
[9] R.A. Hibschweiler and T.H. MacGregor, Multipliers of families of Cauchy-Stieltjes transforms, *Trans. Amer. Math. Soc.* 331 (1992), 377-394.
[10] R.A. Hibschweiler and T.H. MacGregor, Fractional Cauchy Transforms, Chapman and Hall/CRC, Boca Raton, 2006.

Department of Mathematics
Bowdoin College
Brunswick, Maine 04011

Contemporary Mathematics
Volume **454**, 2008

Abstract

Alec L. Matheson* (matheson@math.lamar.edu), Department of Mathematics, Lamar University, Beaumont, TX 77710. *Continuous functions in star-invariant subspaces.* This talk will examine the continuous functions in the star-invariant subspaces of H^p. The existence of such functions was shown by Alexsandrov and further discussed by Dyakonov. This talk will discuss further properties of these continuous functions. (Received February 21, 2006)

Contemporary Mathematics
Volume **454**, 2008

Indestructible Blaschke products

William T. Ross

In memory of Alec L. Matheson

1. Introduction

Consider the following set of linear fractional maps

$$\tau_a(z) := \frac{z-a}{1-\overline{a}z}, \quad |a| < 1.$$

Each τ_a is an automorphism of the open unit disk $\mathbb{D} := \{z \in \mathbb{C} : |z| < 1\}$ and $\tau_a(\partial\mathbb{D}) = \partial\mathbb{D}$. For an inner function ϕ, the *Frostman shifts*

$$\phi_a(z) := \tau_a \circ \phi(z) = \frac{\phi(z) - a}{1 - \overline{a}\phi(z)}$$

are certainly inner functions. A celebrated theorem of Frostman [**9**] says that ϕ_a is actually a Blaschke product for every $|a| < 1$ with the possible exception of a set of logarithmic capacity zero. In this survey paper, we explore the class of Blaschke products for which this exceptional set is empty. These Blaschke products are called *indestructible* and have some intriguing properties.

2. Frostman's theorems

If $(a_n)_{n\geqslant 1}$ is a sequence of points in $\mathbb{D}$, $p \in \mathbb{N} \cup \{0\}$, and $\gamma \in \mathbb{R}$, a necessary and sufficient condition that the infinite product

$$B(z) = e^{i\gamma} z^p \prod_{n=1}^{\infty} \frac{\overline{a_n}}{|a_n|} \frac{a_n - z}{1 - \overline{a_n} z}$$

defines an analytic function on $\mathbb{D}$ is that the series

$$\sum_{n=1}^{\infty} (1 - |a_n|)$$

converges. Such sequences $(a_n)_{n\geqslant 1}$ are called *Blaschke sequences* and the product B is called a *Blaschke product*. The function B is analytic on $\mathbb{D}$, has zeros precisely at the origin and the a_n's (repeated according to their multiplicity), and satisfies

2000 *Mathematics Subject Classification.* Primary 30D50; Secondary 30D35.
Key words and phrases. Blaschke products, Frostman shifts, inner functions.

$|B(z)| < 1$ for all $z \in \mathbb{D}$. Furthermore, by a well-known theorem of Fatou [**3**, Ch. 2] [**8**, Ch. 2], the radial limit function

$$B^*(\zeta) := \lim_{r \to 1^-} B(r\zeta)$$

exists and satisfies $|B^*(\zeta)| = 1$ for almost every $\zeta \in \partial\mathbb{D}$, with respect to (normalized) Lebesgue measure m on $\partial\mathbb{D}$.

REMARK 2.1.
1. In what follows, we use the notation $B^*(\zeta)$ to denote the radial limit value of B at ζ whenever it exists (whether or not it is unimodular).
2. This paper will cover a selection of results about Blaschke products. All the basic properties of Blaschke products, and more, are covered in [**3, 5, 6, 8, 12, 18, 26**].

For a particular point $\zeta \in \partial\mathbb{D}$, there is the following refinement of Fatou's theorem [**10**] (see also [**3**, p. 33]).

THEOREM 2.2 (Frostman). *A necessary and sufficient condition that a Blaschke product B, with zeros $(a_n)_{n \geqslant 1}$, and all its subproducts have radial limits of modulus one at $\zeta \in \partial\mathbb{D}$ is that*

$$\sum_{n=1}^{\infty} \frac{1 - |a_n|}{|\zeta - a_n|} < \infty. \tag{2.3}$$

The Frostman theorems (like Theorem 2.2 above and Theorem 2.13, Theorem 2.14, and Theorem 2.18 below) are not always standard material for many complex analysts and so, for the sake of completeness and to give the reader a sense of how all these ideas are related, we will outline parts of the proofs of his theorems.

In our discussion below, we will only use one direction of Theorem 2.2 so we prove this one direction and point the reader to [**3**, p. 34] for the proof of the other. Suppose, for fixed $\zeta \in \partial\mathbb{D}$, the condition in eq.(2.3) holds. We wish to show that

$$B^*(\zeta) := \lim_{r \to 1^-} B(r\zeta)$$

exists and $|B^*(\zeta)| = 1$. The proof of the same result for any sub-product will follow in a similar way. Without loss of generality, we can assume $\zeta = 1$. First check the following inequalities

$$|1 - \bar{a}r| > 1 - r, \quad |1 - \bar{a}r| > \frac{1}{2}|1 - a|, \quad 0 < r < 1, \quad |a| < 1. \tag{2.4}$$

Second, use induction to verify that for a sequence $(b_n)_{n \geqslant 1} \subset (0,1)$, we have

$$\prod_{n=1}^{N} (1 - b_n) \geqslant 1 - \sum_{n=1}^{N} b_n, \quad \forall N \in \mathbb{N}. \tag{2.5}$$

Third, one can verify, via a routine computation, the identity

$$\frac{|r - \overline{a_n}|^2}{|1 - \overline{a_n} r|^2} = 1 - \frac{(1 - r^2)(1 - |a_n|^2)}{|1 - \overline{a_n} r|^2}$$

and so

$$\begin{aligned}|B(r)|^2 &= \prod_{n=1}^{\infty} \frac{|r - \overline{a_n}|^2}{|1 - \overline{a_n} r|^2} \\ &= \prod_{n=1}^{\infty} \left\{1 - \frac{(1-r^2)(1-|a_n|^2)}{|1-\overline{a_n} r|^2}\right\} \\ &\geqslant 1 - \sum_{n=1}^{\infty} \frac{(1-r^2)(1-|a_n|^2)}{|1-\overline{a_n} r|^2}, \qquad \text{(by eq.(2.5))} \\ &= 1 - \sum_{n=1}^{\infty} \frac{(1-r^2)(1-|a_n|^2)}{|1-\overline{a_n} r||1-\overline{a_n} r|}.\end{aligned}$$

Now use the inequalities in eq.(2.4) and the dominated convergence theorem to get

$$\lim_{r \to 1^-} |B(r)| = 1. \tag{2.6}$$

To finish, we need to show that

$$\lim_{r \to 1^-} \arg B(r)$$

exists. Use the identity

$$|a_n| \frac{\overline{a_n}}{|a_n|} \frac{a_n - r}{1 - \overline{a_n} r} = 1 - \frac{1 - |a_n|^2}{1 - \overline{a_n} r}$$

to get

$$\arg B(r) = \sum_{n=1}^{\infty} \arg \left\{1 - \frac{1-|a_n|^2}{1-\overline{a_n} r}\right\}. \tag{2.7}$$

If $a_n = \alpha_n + i\beta_n$, where $\alpha_n, \beta_n \in \mathbb{R}$, some trigonometry will show that

$$\arg\left(1 - \frac{1-|a_n|^2}{1-\overline{a_n} r}\right) = \sin^{-1}\left(\frac{\beta_n r(1-|a_n|^2)}{|a_n||a_n - r||1-\overline{a_n} r|}\right).$$

From here, one can argue that the right-hand side of eq.(2.7) converges absolutely and uniformly in r and so

$$\lim_{r \to 1^-} \arg B(r)$$

exists. Combine this with eq.(2.6) to complete one direction of the proof. See [**3**, p. 34] for the other direction.

REMARK 2.8. (1) If the zeros $(a_n)_{n \geqslant 1}$ do not accumulate at ζ, the condition in eq.(2.3) is easily satisfied and in fact, B extends analytically to an open neighborhood of ζ [**18**, p. 68].

(2) The zeros can accumulate at ζ and eq.(2.3) can still hold. For example, let $t_n \downarrow 0$ satisfy $\sum_n t_n < \infty$ and let

$$a_n = \frac{1}{2} + \frac{1}{2} e^{it_n}.$$

Notice how these zeros lie on the circle $|z - \frac{1}{2}| = \frac{1}{2}$, which is internally tangent to $\partial\mathbb{D}$ at $\zeta = 1$, and accumulate at $\zeta = 1$. A computation shows that

$$1 - |a_n|^2 = \frac{1}{2}(1 - \cos t_n) \asymp t_n^2, \quad |1 - a_n| = \frac{\sqrt{2}}{2}\sqrt{1 - \cos t_n} \asymp t_n$$

and so

$$\sum_{n=1}^{\infty} \frac{1-|a_n|^2}{|1-a_n|} \asymp \sum_{n=1}^{\infty} t_n < \infty.$$

Notice how this infinite Blaschke product B with zeros $(a_n)_{n\geqslant 1}$ satisfies $|B^*(\zeta)| = 1$ for *every* $\zeta \in \partial\mathbb{D}$.

(3) With more work, one can even arrange the zeros of B to satisfy the much stronger condition

$$\sup_{\zeta \in \partial\mathbb{D}} \sum_{n=1}^{\infty} \frac{1-|a_n|}{|\zeta - a_n|} < \infty.$$

We will get to this in the last section.

(4) So far, we have examined when $B^*(\zeta)$ exists and has modulus one. Frostman [9] showed that the Blaschke product with zeros $a_n = 1-n^{-2}$ satisfies $B^*(1) = 0$.

(5) If $B^*(\zeta)$ exists for *every* $\zeta \in \partial\mathbb{D}$, then results in [1, 5] say that if E is the set of accumulation points of $(a_n)_{n\geqslant 1}$, then (a) E is a closed nowhere dense subset of $\partial\mathbb{D}$, (b) the function $\zeta \to B^*(\zeta)$ is discontinuous at ζ_0 if and only if $\zeta_0 \in E$.

By Fatou's theorem, the radial limit function

$$\phi^*(\zeta) := \lim_{r \to 1^-} \phi(r\zeta),$$

for a bounded analytic function ϕ on $\mathbb{D}$, exists for m-almost every $\zeta \in \partial\mathbb{D}$ [8, p. 6]. If $|\phi^*(\zeta)| = 1$ for almost every ζ, then ϕ is called an *inner function* and can be factored as

$$\phi(z) = \left\{ e^{i\gamma} z^p \prod_{n=1}^{\infty} \frac{\overline{a_n}}{|a_n|} \frac{a_n - z}{1 - \overline{a_n} z} \right\} \exp\left(- \int_{\partial\mathbb{D}} \frac{\zeta + z}{\zeta - z} d\mu(\zeta) \right). \tag{2.9}$$

Here μ is a positive finite measure on $\partial\mathbb{D}$ with $\mu \perp m$. The first factor in eq.(2.9) is the Blaschke factor and is an inner function. The second term in eq.(2.9) is called the *singular inner factor*. By a theorem of Fatou [8, p. 4],

$$\lim_{r \to 1^-} \int_{\partial\mathbb{D}} \frac{1-r^2}{|\zeta - re^{i\theta}|^2} d\mu(\zeta) = (D\mu)(e^{i\theta}) \tag{2.10}$$

whenever $D\mu(e^{i\theta})$, the symmetric derivative of μ at $e^{i\theta}$, exists (and we include the possibility that $(D\mu)(e^{i\theta}) = \infty$). By the Lebesgue differentiation theorem, $D\mu$ exists at m-almost every $e^{i\theta}$. Moreover, since $\mu \perp m$, we know that

$$D\mu = 0 \quad m\text{-a.e.} \quad \text{and} \quad D\mu = \infty \quad \mu\text{-a.e.} \tag{2.11}$$

See [30, p. 156 - 158] for the proofs of eq.(2.11). The first identity in eq.(2.11), along with the identity

$$\left| \exp\left(- \int_{\partial\mathbb{D}} \frac{\zeta + re^{i\theta}}{\zeta - re^{i\theta}} d\mu(\zeta) \right) \right| = \exp\left(- \int_{\partial\mathbb{D}} \frac{1-r^2}{|\zeta - re^{i\theta}|^2} d\mu(\zeta) \right), \tag{2.12}$$

shows that the radial limits of this second factor are unimodular m-almost everywhere and hence this factor is an inner function. Furthermore, if $\mu \not\equiv 0$ (i.e., the inner function ϕ has a non-trivial singular inner factor), we can use the second identity in eq.(2.11) along with eq.(2.12) once again to obtain the following theorem of Frostman [9].

THEOREM 2.13 (Frostman). *If an inner function ϕ has a non-trivial singular inner factor, there is a point $\zeta \in \partial\mathbb{D}$ such that $\phi^*(\zeta) = 0$.*

From Remark 2.8 (4), the condition $\phi^*(\zeta) = 0$ for some $\zeta \in \partial\mathbb{D}$ does not completely determine the presence of a non-trivial inner factor. Another result of Frostman (see [**9**, p. 107] or [**3**, p. 32]) completes the picture.

THEOREM 2.14 (Frostman). *An inner function ϕ is a Blaschke product if and only if*

$$\lim_{r \to 1^-} \int_0^{2\pi} \log|\phi(re^{i\theta})| d\theta = 0. \tag{2.15}$$

Again, for the sake of giving the reader a feel for how all these ideas are related, and since this result will be used later, we outline a proof. We follow [**3**, p. 32]. Indeed, suppose $\phi = B$, a Blaschke product. Let B_n be the product of the first n terms of B and, given $\epsilon > 0$, choose a large n so that

$$\left|\frac{B}{B_n}(0)\right| > 1 - \varepsilon.$$

Thus,

$$\begin{aligned}
0 &\geqslant \lim_{r\to 1^-} \int_{\partial\mathbb{D}} \log|B(r\zeta)| dm(\zeta) \\
&= \lim_{r\to 1^-} \int_{\partial\mathbb{D}} \log\left|\frac{B}{B_n}(r\zeta)\right| dm(\zeta) - \lim_{r\to 1^-} \int_{\partial\mathbb{D}} \log|B_n(r\zeta)| dm(\zeta) \\
&= \lim_{r\to 1^-} \int_{\partial\mathbb{D}} \log\left|\frac{B}{B_n}(r\zeta)\right| dm(\zeta) \\
&\geqslant \log(1-\varepsilon).
\end{aligned}$$

The last inequality comes from the sub-mean value property applied to the subharmonic function $\log|B/B_n|$ [**12**, p. 36]. It follows that eq.(2.15) holds for $\phi = B$.

Now suppose that ϕ is inner and eq.(2.15) holds. Factor $\phi = Be^g$, where

$$g(z) := -\int_{\partial\mathbb{D}} \frac{\zeta + z}{\zeta - z} d\mu(\zeta)$$

and notice, using the fact that $\Re g$ is non-positive and harmonic along with the mean value property for harmonic functions, that if $\Re g$ has a zero in $\mathbb{D}$, then $\Re g \equiv 0$ on $\mathbb{D}$ and consequently $\mu \equiv 0$. Use the mean value property again to see that

$$\int_{\partial\mathbb{D}} \log|\phi(r\zeta)| dm(\zeta) = \int_{\partial\mathbb{D}} \log|B(r\zeta)| dm(\zeta) + \Re g(0).$$

As $r \to 1^-$, the integral on the right-hand side approaches zero since B is a Blaschke product (see above) and the integral on the left-hand side approaches zero by assumption. This means that $\Re g(0) = 0$ and so, by what we said before, $\mu \equiv 0$ and so $\phi = B$ is a Blaschke product. This completes the proof.

The linear fractional maps

$$\tau_a(z) := \frac{z - a}{1 - \bar{a}z}, \quad |a| < 1,$$

are automorphisms of $\mathbb{D}$ (the complete set of automorphisms of $\mathbb{D}$ is $\{\zeta\tau_a : \zeta \in \partial\mathbb{D}, a \in \mathbb{D}\}$) and also satisfy $\tau_a(\partial\mathbb{D}) = \partial\mathbb{D}$. So certainly the Frostman shifts

$$\phi_a := \tau_a \circ \phi, \quad |a| < 1,$$

are all inner functions. However, some of them might not be Blaschke products - even if ϕ is a Blaschke product. For example (see eq.(4.6)) the function

$$B(z) := \tau_{1/2}\left(\exp\left(-\frac{1+z}{1-z}\right)\right)$$

turns out to be a Blaschke product. However,

$$B_{-1/2}(z) := \tau_{-1/2} \circ B(z) = \exp\left(-\frac{1+z}{1-z}\right)$$

is a singular inner function. Define the *exceptional set* $\mathcal{E}(\phi)$ for ϕ to be

$$\mathcal{E}(\phi) := \{a \in \mathbb{D} : \tau_a \circ \phi \text{ is not a Blaschke product}\}. \tag{2.16}$$

This exceptional set $\mathcal{E}(\phi)$ has some very special properties. The first was observed in [**16**] (see also [**22**]).

PROPOSITION 2.17. *The exceptional set $\mathcal{E}(\phi)$ of an inner function ϕ is of type F_σ.*

PROOF. We follow the proof from [**22**, p. 53]. For each $a \in \mathbb{D}$, we the function

$$r \mapsto \int_{\partial\mathbb{D}} \log|\phi_a(r\zeta)|dm(\zeta)$$

is increasing on $[0, 1)$ [**8**, p. 9] and so from Theorem 2.14, we see that $a \in \mathcal{E}(\phi)$ if and only if

$$\lim_{r\to 1^-} \int_{\partial\mathbb{D}} \log|\phi_a(r\zeta)|dm(\zeta) < 0.$$

For $r \in [0, 1)$ and $a \in \mathbb{D}$, let

$$I(r, a) := \int_{\partial\mathbb{D}} \log|\phi_a(r\zeta)|dm(\zeta)$$

and observe how, for fixed r, $I(r, a)$ is a continuous function of a.

For fixed $r \in (0, 1)$ and $k \in \mathbb{N}$, let

$$F(r, k) := \left\{a \in \mathbb{D} : I(r, a) \leqslant -\frac{1}{k}\right\}.$$

Notice that $F(r, k)$ is relatively closed in $\mathbb{D}$. Finally, we observe that

$$\mathcal{E}(\phi) = \bigcup_{k=1}^{\infty} \bigcap_{n=2}^{\infty} F\left(1 - \frac{1}{n}, k\right)$$

which proves the result. □

The exceptional set $\mathcal{E}(\phi)$ satisfies one more special property. In order to explain this, we need the definition of logarithmic capacity. We follow [**12**, p. 78]. For a compact set $K \subset \mathbb{D}$ and positive finite measure supported on K, consider the Green potential

$$G_\sigma(z) := \int \log\left|\frac{1 - \overline{\zeta}z}{\zeta - z}\right| d\sigma(\zeta)$$

and note that

$$0 \leqslant G_\sigma(z) \leqslant \infty, \quad z \in \mathbb{D}.$$

Since K is a compact subset of $\mathbb{D}$, G_σ is continuous near $\partial\mathbb{D}$ and in fact

$$G_\sigma(\zeta) = 0, \quad \zeta \in \partial\mathbb{D}.$$

We will say K has *positive logarithmic capacity* if there is a positive (non-zero) measure σ supported on K such that G_σ is bounded on $\mathbb{D}$. Otherwise, we say that K has *zero logarithmic capacity.* We say a Borel set $E \subset \mathbb{D}$ has positive logarithmic capacity if it contains a compact subset of positive logarithmic capacity. For example, if A denotes two-dimensional Lebesgue area measure in the plane and $A(E) > 0$, a computation shows that G_A is bounded on $\mathbb{D}$. Thus any set of positive area has positive logarithmic capacity. However sets of zero logarithmic capacity are much 'thinner'. For example, Borel subsets of logarithmic capacity zero must have zero area and compact subsets of zero logarithmic capacity must be totally disconnected. There are various other ways to define logarithmic capacity, depending on the particular application. However, they all have the same sets of logarithmic capacity zero. Two excellent sources which sort all this out are [**11, 29**].

This next result of Frostman [**9**] says that $\mathcal{E}(\phi)$ is a small set.

THEOREM 2.18 (Frostman). *For an inner function ϕ, $\mathcal{E}(\phi)$ has logarithmic capacity zero.*

PROOF. Suppose $\mathcal{E}(\phi)$ has positive logarithmic capacity. By Theorem 2.14, there is a compact subset K of positive logarithmic capacity such that

$$\lim_{r \to 1^-} \int_{\partial\mathbb{D}} \log \left| \frac{1 - \overline{w}\phi(r\zeta)}{w - \phi(r\zeta)} \right| dm(\zeta) > 0, \quad \forall w \in K. \tag{2.19}$$

Moreover, by the definition of logarithmic capacity, there is a positive non-zero measure σ supported on K such that G_σ is bounded on $\mathbb{D}$. We than have

$$\begin{aligned}
0 &= \lim_{r \to 1^-} \int_{\partial\mathbb{D}} G_\sigma(\phi(r\zeta)) dm(\zeta) \quad \text{(dominated convergence theorem)} \\
&= \lim_{r \to 1^-} \int_K \left(\int_{\partial\mathbb{D}} \log \left| \frac{1 - \overline{w}\phi(r\zeta)}{w - \phi(r\zeta)} \right| dm(\zeta) \right) d\sigma(w) \quad \text{(Fubini's theorem)} \\
&\geqslant \int_K \left(\varliminf_{r \to 1^-} \int_{\partial\mathbb{D}} \log \left| \frac{1 - \overline{w}\phi(r\zeta)}{w - \phi(r\zeta)} \right| dm(\zeta) \right) d\sigma(w) \quad \text{(Fatou's lemma)} \\
&> 0 \quad \text{(by eq.(2.19))}
\end{aligned}$$

which is a contradiction. □

Let us make a few remarks about the limits of Theorem 2.18.

REMARK 2.20. (1) Frostman [**9**, p. 113] showed that if E is relatively closed in $\mathbb{D}$ and has logarithmic capacity zero, then there is an inner function ϕ with $\mathcal{E}(\phi) = E$ (see also [**3**, p. 37] and the next two comments).

(2) Recall from Proposition 2.17 and Theorem 2.18 that $\mathcal{E}(\phi)$ is an F_σ set of logarithmic capacity zero. The authors in [**22**] showed that if $E \subset \mathbb{D}$ is of type F_σ and has logarithmic capacity zero, then there is an inner function ϕ such that $\mathcal{E}(\phi) = E$.

(3) Suppose that E is a closed subset of $\mathbb{D}$, $0 \notin E$, and E has logarithmic capacity zero. We claim that there is a Blaschke product B such that $B_a := \tau_a \circ B$ is a Blaschke product whenever $a \in \mathbb{D} \setminus E$ and B_a is a singular inner function whenever $a \in E$. To see this, let B be the universal covering map from $\mathbb{D}$ onto $\mathbb{D} \setminus E$ [**7**, p. 125]. Notice that $B^*(\zeta) \in \partial\mathbb{D} \cup E$. First note that B is inner. Indeed, suppose that $|B^*(\zeta)| < 1$ for $\zeta \in A$ and $m(A) > 0$. Then $B^*(A) \subset E$ and, since E has logarithmic capacity zero, we see that $B \equiv 0$ [**3**, p. 37] which is a contradiction. Second, note that B_a is a Blaschke product for all $a \in \mathbb{D} \setminus E$. Indeed, B_a maps $\mathbb{D}$ onto $\mathbb{D} \setminus \tau_a(E)$ and $0 \notin \tau_a(E)$. Moreover, $B_a^*(\zeta) \in \partial\mathbb{D} \cup \tau_a(E)$ and so $B_a^*(\zeta)$ can never be zero. An application of Theorem 2.13 completes the proof. Third, B_a is a singular inner function whenever $a \in E$. To see this, note that B maps $\mathbb{D}$ onto $\mathbb{D} \setminus E$ and so $a \notin B(\mathbb{D})$ which means the inner function B_a has no zeros. Thus B_a must be a singular inner function.

(4) If one is willing to work even harder in the previous example, one can find an *interpolating* Blaschke product B such that B_a is an interpolating Blaschke product for all $a \in \mathbb{D} \setminus E$ while B_a is a singular inner function whenever $a \in E$ [**14**, Theoerm 1.1]. In fact, the above proof is part of this one.

3. Indestructible Blaschke products

From Frostman's theorem (Theorem 2.18), we know that the exceptional set $\mathcal{E}(\phi)$ of an inner function ϕ is small. A Blaschke product B is *indestructible* if $\mathcal{E}(B) = \varnothing$. This next technical result from [**21**] helps show that indestructible Blaschke products actually exist.

PROPOSITION 3.1. *If B is a Blaschke product such that $B^*(\zeta)$ is never equal to $a \in \mathbb{D} \setminus \{0\}$, then B is indestructible.*

PROOF. Suppose that for some $a \in \mathbb{D} \setminus \{0\}$, $B_a = \tau_a \circ B$ has a non-trivial singular inner factor. By Theorem 2.13, there is a $\zeta \in \partial\mathbb{D}$ such that $B_a^*(\zeta) = 0$. However, for $0 < r < 1$,

$$|B_a(r\zeta)| \geqslant \frac{1}{2}|B(r\zeta) - a|$$

and so, taking limits as $r \to 1^-$, we see that $B^*(\zeta) = a$, which contradicts our assumption. □

COROLLARY 3.2. *If B is a Blaschke product whose zeros $(a_n)_{n \geqslant 1}$ satisfy*

$$\sum_{n=1}^{\infty} \frac{1 - |a_n|}{|\zeta - a_n|} < \infty \tag{3.3}$$

for every $\zeta \in \partial\mathbb{D}$, then B is indestructible.

PROOF. By Theorem 2.2, $|B^*(\zeta)| = 1$ for every $\zeta \in \mathbb{T}$. Now apply Proposition 3.1. □

Certainly any finite Blaschke product satisfies eq.(3.3). The infinite Blaschke product in Remark 2.8 (2) also satisfies eq.(3.3) and thus is indestructible.

Let us say a few words about the origins of the concept of indestructibility. The following idea was explored by Heins [**15, 16**] for analytic functions on Riemann surfaces but, for the sake of simplicity, we outline this idea when the Riemann surface is the unit disk. Our discussion has not only historical value, but will be useful when we discuss a fascinating example of Morse later on.

If $f : \mathbb{D} \to \mathbb{D}$ is analytic and $a \in \mathbb{D}$, the function $z \mapsto -\log|f_a(z)|$, where $f_a = \tau_a \circ f$, is superharmonic on $\mathbb{D}$ (i.e., $\log|f_a|$ is subharmonic on $\mathbb{D}$). Using the classical inner-outer factorization theorem [**8**, Ch. 2], one can show that

$$-\log|f_a(z)| = \sum_{f(w)=a} -n(w)\log|\tau_w(z)| + u_a(z), \tag{3.4}$$

where $n(w)$ is the multiplicity of the zero of $f(z) - a$ at $z = w$, and u_a is a non-negative harmonic function on $\mathbb{D}$. The focus of Heins' work is the residual term u_a. His first observation is that u_a is the greatest harmonic minorant of $-\log|f_a|$. Moreover, since u_a is a non-negative harmonic function on $\mathbb{D}$, Herglotz's theorem [**8**, p. 2] yields a positive measure μ_a on $\partial\mathbb{D}$ such that

$$u_a(z) = (P\mu_a)(z) = \int_{\partial\mathbb{D}} \frac{1-|z|^2}{|\zeta - z|^2} d\mu_a(\zeta),$$

the Poisson integral of μ_a. Heins proves that if $\mu_a = \nu_a + \sigma_a$ is the Lebesgue decomposition of μ_a, where $\nu_a \ll m$ and $\sigma_a \perp m$, then the m-almost everywhere defined function

$$q_a(\zeta) := \log\left|\frac{1-\bar{a}f^*(\zeta)}{a - f^*(\zeta)}\right|$$

is integrable on $\partial\mathbb{D}$ and

$$d\nu_a = q_a dm. \tag{3.5}$$

In the general setting, and the actual focus of his work, Heins examines the residual term u_a in Lindelöf's theorem

$$G_{S_1}(f(z), a) = \sum_{f(w)=a} n(w) G_{S_2}(z, w) + u_a(z),$$

where S_1 and S_2 are Riemann surfaces with positive ideal boundary, f is a conformal map from S_1 to S_2, and G_{S_j} is the Green's function for S_j. To study the residual term u_a in this general setting, Herglotz's theorem and the Lebesgue decomposition theorem are replaced by an old decomposition theorem of Parreau [**27**, Théorème 12] (see also [**17**, p. 7]). When $S_1 = S_2 = \mathbb{D}$, observe that

$$G_{S_j}(z, a) = -\log\left|\frac{z-a}{1-\bar{a}z}\right|.$$

We state this next theorem in the special case of the disk but refer the reader to Heins' paper where an analog of this theorem holds for Riemann surfaces.

THEOREM 3.6 (Heins). *The functions u_a and $P\nu_a$ satisfy the following properties.*

1. *Either $P\nu_a(z) = 0$ for all $(a, z) \in \mathbb{D} \times \mathbb{D}$ or $P\nu_a(z) \neq 0$ for all $(a, z) \in \mathbb{D} \times \mathbb{D}$.*
2. *The set $\{a \in \mathbb{D} : u_a - P\nu_a > 0\}$ is an F_σ set of logarithmic capacity zero.*

PROOF. Observe that if $a \in \mathbb{D}$ is fixed and $P\nu_a(z) = 0$ for some $z \in \mathbb{D}$, we can use the fact that $P\nu_a$ is a non-negative harmonic function along with the mean value property of harmonic functions to argue that $P\nu_a \equiv 0$. Thus, from eq.(3.5), we have, this particular a,

$$0 = \lim_{r \to 1^-} P\nu_a(r\zeta) = \log \left| \frac{1 - \overline{a} f^*(\zeta)}{a - f^*(\zeta)} \right|, \quad \text{a.e. } \zeta \in \partial\mathbb{D}.$$

Whence it follows that $|f^*(\zeta)| = 1$ almost everywhere, i.e., f is inner. The fact that f is inner along with the fact that τ_b maps $\partial\mathbb{D}$ to $\partial\mathbb{D}$ for each $b \in \mathbb{D}$ shows that

$$\lim_{r \to 1^-} P\nu_b(\zeta) = 0 \quad \text{a.e. } \zeta \in \partial\mathbb{D}.$$

Thus, from eq.(2.10), we see that for each $b \in \mathbb{D}$,

$$0 = \lim_{r \to 1^-} P\nu_b(\zeta) = D\nu_b(\zeta) \quad \text{a.e. } \zeta \in \partial\mathbb{D}$$

and so $\nu_b \perp m$. But since $\nu_b \ll m$ it must be the case that $\nu_b \equiv 0$. Thus we have shown part (1) of the theorem.

To avoid some technicalities, and to keep our focus on Blaschke products, let us prove part (2) of the theorem in the special case when $P\nu_a \equiv 0$ for some (equivalently all) a. Note that f is inner. If u_a has a zero in $\mathbb{D}$, then, as argued before using the mean value property of harmonic functions, $u_a \equiv 0$. Recall from our earlier discussion that

$$u_a = \mathfrak{m}(-\log|f_a|),$$

where $\mathfrak{m}$ denotes the greatest harmonic minorant. If we factor $f_a = bg$ as the product of a Blaschke product b and a singular inner function g, one can argue that

$$\mathfrak{m}(-\log|f_a|) = \mathfrak{m}(-\log|b|) + \mathfrak{m}(-\log|g|).$$

It follows from Theorem 2.14 and a technical fact about greatest harmonic minorants [**12**, p. 38], that $\mathfrak{m}(-\log|b|) \equiv 0$. But since we are assuming $u_a \equiv 0$, we have $\mathfrak{m}(-\log|g|) \equiv 0$. However, g has no zeros in $\mathbb{D}$ and so $-\log|g|$ is a non-negative harmonic function on $\mathbb{D}$ and thus

$$0 \equiv \mathfrak{m}(-\log|g|) = -\log|g|.$$

Hence $g \equiv e^{ic}$, $c \in \mathbb{R}$, equivalently, f_a is a Blaschke product. Thus we have shown $u_a \equiv 0 \Rightarrow f_a$ is a Blaschke product. If f_a is a Blaschke product, then, as pointed out before, $u_a = \mathfrak{m}(-\log|f_a|) \equiv 0$. It follows that

$$\{a \in \mathbb{D} : u_a > 0\} = \mathcal{E}(f). \tag{3.7}$$

Now use Proposition 2.17 and Theorem 2.18. □

In summary, $u_a \equiv 0$ if and only if f_a is a Blaschke product. Moreover, $u_a \equiv 0$ for every $a \in \mathbb{D}$ if and only if f is an indestructible Blaschke product. Heins did not coin the term 'indestructible' in his work. McLaughlin [**21**] was the first to use this term and to explore the properties of these products.

4. Zeros of indestructible Blaschke products

McLaughlin [**21**] determined a characterization of the indestructible Blaschke products in terms of their level sets. Suppose ϕ is inner and $a \in \mathbb{D} \setminus \{\phi(0)\}$. Let $(w_j)_{j \geqslant 1}$ be the solutions to $\phi(z) - a = 0$ and factor

$$\phi_a = \frac{\phi - a}{1 - \overline{a}\phi} = b \cdot s,$$

where b is a Blaschke product whose zeros are $(w_j)_{j\geqslant 1}$ and s is a singular inner function. Taking absolute values of both sides of the above equation and evaluating at $z = 0$, we get

$$\left|\frac{\phi(0) - a}{1 - \overline{a}\phi(0)}\right| = \left(\prod_{j=1}^{\infty} |w_j|\right) |s(0)|.$$

As discussed in the proof of Theorem 2.14, notice that $|s(0)| = 1$ if and only if s is a unimodular constant, i.e., ϕ_a is a Blaschke product. In other words, for $a \neq \phi(0)$, ϕ_a is a Blaschke product if and only if

$$\left|\frac{\phi(0) - a}{1 - \overline{a}\phi(0)}\right| = \prod_{j=1}^{\infty} |w_j|.$$

What happens when $a = \phi(0)$? Let

$$\phi(z) - \phi(0) = b_n z^n + b_{n+1} z^{n+1} + \cdots$$

be the Taylor series of $\phi - \phi(0)$ about $z = 0$ and let $(z_j)_{j\geqslant 1}$ be the non-zero zeros of $\phi(z) - \phi(0)$. As before, write

$$\frac{1}{z^n} \frac{\phi - \phi(0)}{1 - \overline{\phi(0)}\phi} = b \cdot s,$$

where s is a singular inner function and b is the Blaschke product whose zeros are $(z_j)_{j\geqslant 1}$. Again, take absolute values of both sides of the above expression and evaluate at $z = 0$ to get

$$\frac{|b_n|}{1 - |\phi(0)|^2} = \left(\prod_{j=1}^{\infty} |z_j|\right) |s(0)|.$$

Moreover, $\phi = \phi_0$ is a Blaschke product if and only if

$$\frac{|b_n|}{1 - |\phi(0)|^2} = \prod_{j=1}^{\infty} |z_j|.$$

Combining these observations, we have shown the following theorem.

THEOREM 4.1 (McLaughlin). *Using the notation above, a Blaschke product B is indestructible if and only if*

$$\left|\frac{B(0) - a}{1 - \overline{a}B(0)}\right| = \prod_{j=1}^{\infty} |w_j|, \quad \forall a \neq B(0),$$

and

$$\frac{|b_n|}{1 - |\phi(0)|^2} = \prod_{j=1}^{\infty} |z_j|.$$

Though the above theorem gives necessary and sufficient conditions (in terms of the level sets of B) to be indestructible, characterizing indestructibility just in terms of the *zeros* of B seems almost impossible. Consider the following theorem of Morse [**23**].

THEOREM 4.2 (Morse). *There is a Blaschke product B for which $\mathcal{E}(B) \neq \varnothing$ but such that if c is any zero of B, then $\mathcal{E}(B/\tau_c) = \varnothing$.*

In other words, there are 'destructible' Blaschke products which become indestructible when one of their zeros are removed. We will not give all of the technical details here since they are done thoroughly in Morse's paper. However, since they do relate directly to the earlier work of Heins, from the previous section, we will give an outline of Morse's theorem.

Suppose B is a Blaschke product such that the set

$$\{\zeta \in \partial\mathbb{D} : |B^*(\zeta)| < 1\}$$

is at most countable. For $a \in \mathbb{D}$, let

$$u_a := \mathfrak{m}(-\log|B_a|),$$

be the greatest harmonic minorant of the non-negative superharmonic function $-\log|B_a|$. This function is the residual function covered in the previous section (see eq.(3.4)). Since u_a is a non-negative harmonic function on $\mathbb{D}$, Herglotz's theorem says that

$$u_a = P\mu_a,$$

the Poisson integral of a measure μ_a on $\partial\mathbb{D}$. Moreover, since $\log|B_a^*(\zeta)| = 0$ for m-almost every ζ, it follows that $u_a^*(\zeta) = 0$ m-almost everywhere. By Fatou's theorem (see eq.(2.10)),

$$u_a^*(\zeta) = (D\mu_a)(\zeta)$$

at every point where $(D\mu_a)(\zeta)$ exists (and we count the possibility that $(D\mu_a)(\zeta)$ might be equal to $+\infty$). We see two things from this. First, $(D\mu_a)(\zeta) = 0$ for m-almost every ζ and so, by the Lebesgue decomposition theorem, $\mu_a \perp m$. Second, since we are assuming that $\{\zeta \in \partial\mathbb{D} : |B^*(\zeta)| < 1\}$ is at most countable, we can use the facts that

$$\{\zeta \in \partial\mathbb{D} : (D\mu_a)(\zeta) = +\infty\} = \{\zeta : u_a^*(\zeta) = +\infty\} \subset \{\zeta : |B^*(\zeta)| < 1\}$$

and $\{\zeta : (D\mu_a)(\zeta) = +\infty\}$ is a carrier for μ_a (since $\mu_a \perp m$) [**30**, p. 158] to see that μ_a is a discrete measure. It might be the case that $\mu_a \equiv 0$, i.e., B_a is a Blaschke product (see eq.(3.7)).

If we make the further assumption that not only is $\{\zeta : |B^*(\zeta)| < 1\}$ at most countable but B is also destructible, i.e., B_a is not a Blaschke product for some $a \in \mathbb{D}$, we see (see eq.(3.7)) that $u_a > 0$ and so, for this particular a, the discrete measure μ_a above is not identically zero. Define $Q(B)$ to be the union of the carriers of the measures $\{\mu_a : u_a = P\mu_a > 0\}$. Notice that

$$Q(B) \subset \{\zeta \in \partial\mathbb{D} : |B^*(\zeta)| < 1\}, \tag{4.3}$$

and hence is at most countable, and that $Q(B)$ is contained in the accumulation points of the zeros of B. We also see in this case that B is destructible if and only if $Q(B) \neq \varnothing$.

A technical theorem of Morse [**23**, Proposition 3.2] says that if $\zeta \in Q(B)$, then there is an inner function g, a point $a \in \mathbb{D}$, and a $\beta > 0$ such that

$$B_a(z) = g(z)\exp\left(-\beta\frac{\zeta + z}{\zeta - z}\right).$$

Morse says in this case that B is *exponentially destructible* at ζ. It follows from here that for some $\alpha > 0$

$$|B(r\zeta) - a| = O(e^{\frac{-\alpha}{1-r}}), \quad r \to 1^-. \tag{4.4}$$

An argument using this growth estimate (see [**23**, Proposition 3.4]) shows that if c is any zero of B, then

$$Q(B) \cap Q(B/\tau_c) = \varnothing. \tag{4.5}$$

Morse gives a treatment of exponentially destructible Blaschke products beyond what we cover here.

We are now ready to discuss Morse's example. Choose $a \in \mathbb{D} \setminus \{0\}$ and define

$$B(z) := \tau_a \left(\exp\left(-\frac{1+z}{1-z} \right) \right). \tag{4.6}$$

One can see that B is an inner function, $B^*(\zeta)$ exists for every $\zeta \in \partial\mathbb{D}$, and

$$|B^*(\zeta)| = \begin{cases} 1, & \text{if } \zeta \in \partial\mathbb{D} \setminus \{1\}; \\ |a|, & \text{if } \zeta = 1. \end{cases}$$

By Theorem 2.13, B is a Blaschke product. It is also the case, by direct computation, that the zeros of B can only accumulate at $\zeta = 1$. Finally, notice from eq.(4.3) and the identity

$$B_{-a}(z) = \exp\left(-\frac{1+z}{1-z} \right),$$

that

$$Q(B) = \{1\}$$

and so B is destructible, in fact exponentially destructible at 1.

We claim that if c is a zero of B, then B/τ_c (B with the zero at c divided out) is indestructible. Indeed, since

$$\{\zeta : |(B/\tau_c)^*(\zeta)| < 1\} = \{\zeta : |B^*(\zeta)| < 1\} = \{1\}$$

we can apply eq.(4.3) to get

$$Q(B/\tau_c) \subset \{1\}.$$

However, from eq.(4.5) we see that $Q(B/\tau_c) = \varnothing$ which means, from our discussion above, that B/τ_c is indestructible.

5. Classes of indestructible Blaschke products

So far, we have discussed conditions on a Blaschke product that make it indestructible. We now examine a refinement of this question. Suppose that $\mathcal{B}$ is a particular class of Blaschke products and $B \in \mathcal{B}$. What extra assumptions are required of B so that $B_a \in \mathcal{B}$ for all $a \in \mathbb{D}$?

We focus on the class (and certain sub-classes) of $\mathcal{C}$, the *Carleson-Newman* Blaschke products. These are Blaschke products B whose zeros $(a_n)_{n \geqslant 1}$ satisfy the so-called 'conformal invariant' version of the Blaschke condition

$$\sum_{n=1}^{\infty} (1 - |a_n|) < \infty,$$

i.e.,

$$\sup \left\{ \sum_{n=1}^{\infty} (1 - |\varphi(a_n)|) : \varphi \in \mathrm{Aut}(\mathbb{D}) \right\} < \infty.$$

There are several equivalent definitions of $\mathcal{C}$. For example, $B \in \mathcal{C} \Leftrightarrow B$ is the finite product of interpolating Blaschke products $\Leftrightarrow$ the measure $\sum_n (1 - |a_n|^2)\delta_{a_n}$ is a Carleson measure. The standard reference for this is [**12**] but another nice exposition with further references is [**24**, Theorem 2.2]. Two important examples of Blaschke products which belong to $\mathcal{C}$ are $\mathcal{T}$, the *thin* Blaschke products B which satisfy the condition

$$\lim_{n\to\infty}(1-|a_n|^2)|B'(a_n)|=1,$$

and $\mathcal{F}$, the *Frostman* Blaschke products which satisfy the condition

$$\sup_{\zeta\in\partial\mathbb{D}}\sum_{n=1}^{\infty}\frac{1-|a_n|}{|\zeta-a_n|}<\infty.$$

An example of a thin Blaschke product is one whose zeros $(a_n)_{n\geqslant 1}$ satisfy

$$\lim_{n\to\infty}\frac{1-|a_{n+1}|}{1-|a_n|}=0$$

[**13**, Prop. 1.1], while an example of a Frostman Blaschke product is one with zeros $a_n=r_ne^{i\theta_n}$, where $(r_n)_{n\geqslant 1}\subset(0,1)$, $(\theta_n)_{n\geqslant 1}\subset(0,1)$,

$$\sup\left\{\frac{\theta_{n+1}}{\theta_n}:n\geqslant 1\right\}<1 \quad\text{and}\quad \sum_{n=1}^{\infty}\frac{1-r_n}{\theta_n}<\infty$$

[**2**, p. 130].

The thin Blaschke products relate to Douglas algebras and the structure of the bounded analytic functions [**31**] as well as composition operators on the Bloch space [**4**]. The Frostman Blaschke products turn out to be the only inner multipliers of the space of Cauchy transforms of measures on $\partial\mathbb{D}$ [**19**] (see also [**2**]).

Following the definition of $\mathcal{E}(B)$, the exceptional set of a Blaschke product in eq.(2.16), define

$$\mathcal{E}_{\mathcal{C}}(B):=\{a\in\mathbb{D}:B_a\notin\mathcal{C}\};$$
$$\mathcal{E}_{\mathcal{T}}(B):=\{a\in\mathbb{D}:B_a\notin\mathcal{T}\};$$
$$\mathcal{E}_{\mathcal{F}}(B):=\{a\in\mathbb{D}:B_a\notin\mathcal{F}\}.$$

Gorkin and Mortini [**14**, Lemma 3.2] use a result of Tolokonnikov [**31**, p. 884] to show the following.

THEOREM 5.1. *If $B\in\mathcal{T}$ then $\mathcal{E}_{\mathcal{T}}(B)=\varnothing$.*

The current author and Matheson [**20**] use the theory of inner multipliers for the space of Cauchy transforms and the ideas of Tolokonnikov [**31**] and Pekarskiĭ [**28**] to prove the following.

THEOREM 5.2. *If $B\in\mathcal{F}$, then $\mathcal{E}_{\mathcal{F}}(B)=\varnothing$.*

Nicolau [**25**] states necessary and sufficient conditions, in terms of the zeros of B, so that $\mathcal{E}_{\mathcal{C}}(B)=\varnothing$ - which are a bit technical to get into here. We do point out the following.

THEOREM 5.3. (1) *If B is a Blaschke product, then $\mathcal{E}_{\mathcal{C}}(B)$ is closed in $\mathbb{D}$.*
(2) *Given any $0<s<1$, there is a $B\in\mathcal{C}$ such that $\mathcal{E}_{\mathcal{C}}(B)=\{s\leqslant|z|<1\}$.*

REMARK 5.4. (1) Compare this to $\mathcal{E}(B)$ which is an F_σ set of logarithmic capacity zero (Proposition 2.17 and Theorem 2.18).
(2) The first result of the above theorem is contained in [**25**, Lemma 1]. A version of the second result is found in [**25**, §3]. See [**14**, Theorem 4.2] for the version we state here.

There is a sizable literature of deep results which relate the class of Blaschke products

$$\mathcal{P} := \{B : \mathcal{E}_{\mathcal{C}}(B) = \varnothing\}$$

to many ideas in function algebras. We refer the reader to [**24**, p. 287] for a discussion of this and for the exact references.

So far we have discussed when a Blaschke product has the property that all its Frostman shifts belong to a certain class of Blaschke products. We point out two papers [**14, 24**] which discuss when an inner function ϕ (not necessarily a Blaschke product) has the property that ϕ_a belongs to a certain class of Blaschke products (the class $\mathcal{C}$ for example) for all $a \in \mathbb{D} \setminus \{0\}$.

Acknowledgement: The author wishes to thank Raymond Mortini for his suggestions and corrections.

References

1. G. T. Cargo, *A theorem concerning plane point sets with an application to function theory*, J. London Math. Soc. **37** (1962), 169–175. MR 0137838 (25 #1287)
2. J. A. Cima, A. L. Matheson, and W. T. Ross, *The Cauchy transform*, Mathematical Surveys and Monographs, vol. 125, American Mathematical Society, Providence, RI, 2006. MR 2215991 (2006m:30003)
3. E. F. Collingwood and A. J. Lohwater, *The theory of cluster sets*, Cambridge Tracts in Mathematics and Mathematical Physics, No. 56, Cambridge University Press, Cambridge, 1966. MR 0231999 (38 #325)
4. F. Colonna, *Characterisation of the isometric composition operators on the Bloch space*, Bull. Austral. Math. Soc. **72** (2005), no. 2, 283–290. MR 2183409 (2006g:30051)
5. P. Colwell, *On the boundary behavior of Blaschke products in the unit disk*, Proc. Amer. Math. Soc. **17** (1966), 582–587. MR 0193243 (33 #1463)
6. ———, *Blaschke products*, University of Michigan Press, Ann Arbor, MI, 1985, Bounded analytic functions. MR 779463 (86f:30033)
7. J. B. Conway, *Functions of one complex variable. II*, Graduate Texts in Mathematics, vol. 159, Springer-Verlag, New York, 1995. MR 1344449 (96i:30001)
8. P. L. Duren, *Theory of H^p spaces*, Academic Press, New York, 1970. MR 42 #3552
9. O. Frostman, *Potential d'équilibre et capacité des ensembles avec quelques applications à la théorie des fonctions*, Ph.D. thesis, Lund, 1935.
10. ———, *Sur les produits de Blaschke*, Kungl. Fysiografiska Sällskapets i Lund Förhandlingar [Proc. Roy. Physiog. Soc. Lund] **12** (1942), no. 15, 169–182. MR 6,262e
11. J. Garnett and D. Marshall, *Harmonic measure*, New Mathematical Monographs, vol. 2, Cambridge University Press, Cambridge, 2005. MR 2150803 (2006g:31002)
12. J. B. Garnett, *Bounded analytic functions*, Academic Press Inc., New York, 1981. MR 83g:30037
13. P. Gorkin and R. Mortini, *Universal Blaschke products*, Math. Proc. Cambridge Philos. Soc. **136** (2004), no. 1, 175–184. MR MR2034021 (2004m:30056)
14. ———, *Value distribution of interpolating Blaschke products*, J. London Math. Soc. (2) **72** (2005), no. 1, 151–168. MR 2145733 (2005m:30037)
15. M. Heins, *Studies in the conformal mapping of Riemann surfaces. I*, Proc. Nat. Acad. Sci. U. S. A. **39** (1953), 322–324. MR MR0054049 (14,862b)
16. ———, *On the Lindelöf principle*, Ann. of Math. (2) **61** (1955), 440–473. MR 0069275 (16,1011g)
17. ———, *Hardy classes on Riemann surfaces*, Lecture Notes in Mathematics, No. 98, Springer-Verlag, Berlin, 1969. MR 0247069 (40 #338)
18. K. Hoffman, *Banach spaces of analytic functions*, Dover Publications Inc., New York, 1988, Reprint of the 1962 original. MR 1102893 (92d:46066)
19. S. V. Hruščev and S. A. Vinogradov, *Inner functions and multipliers of Cauchy type integrals*, Ark. Mat. **19** (1981), no. 1, 23–42. MR 83c:30027

20. A. Matheson and W. T. Ross, *An observation about Frostman shifts*, Comp. Math. Funct. Thry. **7** (2007), 111 – 126.
21. R. McLaughlin, *Exceptional sets for inner functions*, J. London Math. Soc. (2) **4** (1972), 696–700. MR 0296309 (45 #5370)
22. R. McLaughlin and G. Piranian, *The exceptional set of an inner function*, Österreich. Akad. Wiss. Math.-Naturwiss. Kl. S.-B. II **185** (1976), no. 1-3, 51–54. MR 0447585 (56 #5895)
23. H. S. Morse, *Destructible and indestructible Blaschke products*, Trans. Amer. Math. Soc. **257** (1980), no. 1, 247–253. MR 549165 (80k:30034)
24. R. Mortini and A. Nicolau, *Frostman shifts of inner functions*, J. Anal. Math. **92** (2004), 285–326. MR 2072750 (2005e:30088)
25. A. Nicolau, *Finite products of interpolating Blaschke products*, J. London Math. Soc. (2) **50** (1994), no. 3, 520–531. MR 1299455 (95k:30072)
26. K. Noshiro, *Cluster sets*, Ergebnisse der Mathematik und ihrer Grenzgebiete. N. F., Heft 28, Springer-Verlag, Berlin, 1960. MR 0133464 (24 #A3295)
27. M. Parreau, *Sur les moyennes des fonctions harmoniques et analytiques et la classification des surfaces de Riemann*, Ann. Inst. Fourier Grenoble **3** (1951), 103–197 (1952). MR 0050023 (14,263c)
28. A. A. Pekarskiĭ, *Estimates of the derivative of a Cauchy-type integral with meromorphic density and their applications*, Mat. Zametki **31** (1982), no. 3, 389–402, 474. MR 652843 (83e:30047)
29. T. Ransford, *Potential theory in the complex plane*, London Mathematical Society Student Texts, vol. 28, Cambridge University Press, Cambridge, 1995. MR 1334766 (96e:31001)
30. W. Rudin, *Real and complex analysis*, McGraw-Hill Book Co., New York, 1966. MR 0210528 (35 #1420)
31. V. A. Tolokonnikov, *Carleson's Blaschke products and Douglas algebras*, Algebra i Analiz **3** (1991), no. 4, 186–197. MR 1152609 (93c:46098)

Department of Mathematics and Computer Science, University of Richmond, Richmond, Virginia 23173

E-mail address: wross@richmond.edu

Contemporary Mathematics
Volume **454**, 2008

On Taylor Coefficients and Multipliers in Fock Spaces

James Tung

ABSTRACT. Several necessary or sufficient conditions for a function to be a member of the Fock space F^p_α are proved, which improve previously known results. We also give conditions for a sequence of complex numbers to be a coefficient multipliers from the Fock space F^1_α to ℓ^1.

1. Introduction

For $0 < p < \infty$ and a function f analytic in either the complex plane $\mathbb{C}$ or the unit disk $\mathbb{D}$ in $\mathbb{C}$, the *integral means* are

$$M_p(r, f) = \left(\frac{1}{2\pi}\int_0^{2\pi} \left|f(re^{i\theta})\right|^p \, d\theta\right)^{\frac{1}{p}}.$$

For $\alpha > 0$, the *Fock space* F^p_α consists of all entire functions f such that

$$\|f\|_{F^p_\alpha} = \left(\frac{\alpha p}{2\pi}\int_{\mathbb{C}} \left|f(z)e^{-\frac{\alpha}{2}|z|^2}\right|^p \, dA(z)\right)^{\frac{1}{p}} < \infty.$$

The following theorems are known to hold for functions in F^p_α [**10**].

THEOREM A (Hausdorff–Young Theorem for F^p_α). *Let $1 \le p \le \infty$, let p' be its conjugate index, and let $f(z) = \sum_{n=0}^\infty a_n z^n$ be an entire function.*

(i) For $1 < p \le 2$,

$$f \in F^p_\alpha \Longrightarrow \sum_{n=1}^\infty |a_n|^{p'} \left(\frac{n!}{\alpha^n}\right)^{\frac{p'}{2}} n^{\frac{p'}{4}-\frac{1}{2}} < \infty.$$

For the case $p = 1$, we have

$$f \in F^1_\alpha \Longrightarrow \sup_{n\in\mathbb{N}} |a_n|\sqrt{\frac{n!}{\alpha^n}}\, n^{\frac{1}{4}} < \infty.$$

(ii) For $2 \le p \le \infty$,

$$\sum_{n=1}^\infty |a_n|^{p'} \left(\frac{n!}{\alpha^n}\right)^{\frac{p'}{2}} n^{\frac{p'}{4}-\frac{1}{2}} < \infty \Longrightarrow f \in F^p_\alpha.$$

2000 *Mathematics Subject Classification.* Primary 30B10, 30D15.
Key words and phrases. Fock space, entire function, Taylor coefficients.

THEOREM B (Hardy–Littlewood Theorem for F^p_α). *Let $f(z) = \sum_{n=0}^{\infty} a_n z^n$ be an entire function, and fix $\alpha > 0$.*

(i) For $0 < p \le 2$,

$$\sum_{n=0}^{\infty} |a_n|^p \left(\frac{n!}{\alpha^n}\right)^{\frac{p}{2}} n^{-\frac{p}{4}+\frac{1}{2}} < \infty \Rightarrow f \in F^p_\alpha \Rightarrow \sum_{n=1}^{\infty} |a_n|^p \left(\frac{n!}{\alpha^n}\right)^{\frac{p}{2}} n^{\frac{3p}{4}-\frac{3}{2}} < \infty.$$

(ii) For $2 \le p < \infty$,

$$\sum_{n=0}^{\infty} |a_n|^p \left(\frac{n!}{\alpha^n}\right)^{\frac{p}{2}} n^{\frac{3p}{4}-\frac{3}{2}} < \infty \Rightarrow f \in F^p_\alpha \Rightarrow \sum_{n=1}^{\infty} |a_n|^p \left(\frac{n!}{\alpha^n}\right)^{\frac{p}{2}} n^{-\frac{p}{4}+\frac{1}{2}} < \infty.$$

Note that the conditions involving the sums in Theorems A and B can be restated in terms of membership in various ℓ^p spaces.

In this paper we show that the theorems above can be improved in several ways. First we restrict our attention to functions with *lacunary* Taylor series; that is, those functions of the form $f(z) = \sum_{k=1}^{\infty} a_k z^{n_k}$, where there is a number $\lambda > 1$ such that for every positive integer k, $n_{k+1}/n_k \ge \lambda$. The sequence $\{n_k\}$ of positive integers is called a *lacunary sequence.* We show that a necessary and sufficient condition for membership in F^p_α can be obtained for functions with lacunary Taylor series. Some results along this line have been obtained by Blasco and Galbis [**2**].

THEOREM C ([**2**], Theorem 2.3, 2.5). *Let $\{n_k\}$ be a lacunary sequence. Then for any sequences $\{a_k\}$,*

$$\sum_{k=1}^{\infty} a_k z^{n_k} \in F^1_2 \iff \sum_{k=1}^{\infty} |a_k| \sqrt{\frac{n_k!}{2^{n_k}}}\, n_k^{\frac{1}{4}} < \infty,$$

$$\sum_{k=1}^{\infty} a_k z^{n_k} \in F^\infty_2 \iff \sup_{k\in\mathbb{N}} |a_k| \sqrt{\frac{n_k!}{2^{n_k}}}\, n_k^{-\frac{1}{4}} < \infty.$$

We will extend the result of Blasco and Galbis to F^p_α, for $\alpha > 0$ and $1 \le p < \infty$. This will give explicit examples of functions in F^p_α. As a side note, we give a new proof for an analogous result concerning functions with lacunary Taylor series in the Bergman spaces A^p.

Another way that we can refine Theorems A and B is to make use of the mixed norm sequence spaces: let $0 < p, q < \infty$, and let $\{n_k\}$ be a lacunary sequence. For each positive integer k, let $I_k = [n_k, n_{k+1}) \cap \mathbb{N}$. A sequence of complex numbers $\{a_n\}$ is said to belong to $\ell^{p,q}$ if

$$\|\{a_n\}\|_{\ell^{p,q}} = \left(\sum_{k=1}^{\infty} \left(\sum_{n\in I_k} |a_n|^p\right)^{\frac{q}{p}}\right)^{\frac{1}{q}} < \infty. \tag{1}$$

In the case of $p = \infty$ or $q = \infty$, the respective norms are

$$\|\{a_n\}\|_{\ell^{\infty,q}} = \left(\sum_{k=1}^{\infty} \left(\sup_{n\in I_k} |a_n|\right)^{q}\right)^{\frac{1}{q}},$$

$$\|\{a_n\}\|_{\ell^{p,\infty}} = \sup_{k\in\mathbb{N}} \left(\sum_{n\in I_k} |a_n|^p\right)^{\frac{1}{p}}.$$

We can view $\ell^{p,q}$ as the vector-valued space $L^q(\mathbb{N}, \nu; \ell^p)$ consisting of functions from $\mathbb{N}$ to ℓ^p where ν is the counting measure. We note that $\ell^{p,p}$ is the usual ℓ^p space, but in general $\ell^{p,q}$ and ℓ^p are different spaces. For example, the sequence whose sum on I_k equals $1/k$ is in $\ell^{1,q}$ for any $q > 1$, but not in ℓ^1. Also, since all sequences in ℓ^p and $\ell^{p,q}$ are bounded, it can be shown that the following inclusions hold for $p < q$:

$$\ell^p \subset \ell^{p,q} \subset \ell^q, \qquad \ell^p \subset \ell^{q,p} \subset \ell^q.$$

The $\ell^{p,q}$ spaces are dependent on the choice of the lacunary series $\{n_k\}$, but this will not affect the statements of our theorems. These spaces were introduced by Kellogg [**8**] in his improvements for the classical Hausdorff–Young Theorem. We will similarly improve Theorems A by giving necessary or sufficient conditions in terms of these $\ell^{p,q}$ spaces.

For $0 < p \leq \infty$, and a space of analytic functions X, we say that a sequence of complex numbers $\{\lambda_n\}$ is a *coefficient multiplier* from X to ℓ^p if for every $f(z) = \sum_{n=0}^{\infty} a_n z^n$ in X, we have $\{\lambda_n a_n\} \in \ell^p$; we use the usual notation (X, ℓ^p) for the space of all coefficient multipliers from X to ℓ^p. Coefficient multipliers in H^p and A^p have been well investigated. For example, a characterization of coefficient multipliers from the Hardy space H^1 to ℓ^1 is found in Theorem 6.8 in [**5**]. We also have the following result characterizing (A^1, ℓ^1) due to Blasco (see Theorem 5.1, [**1**]):

THEOREM D. *A sequence of complex numbers $\{\lambda_n\}$ is a multiplier from A^1 to ℓ^1 if and only if $\{n\lambda_n\}$ is in $\ell^{1,\infty}$.*

We will prove a sufficient condition for $\{\lambda_n\}$ to be a multiplier from F^1_α to ℓ^1 and show that it is in some sense the best possible. In the converse direction, we will prove a necessary condition that, curiously, differs from the sufficient condition by a factor of $\sqrt{n}$.

In the proof of theorems, we shall abuse notations and use c and C to represent positive constants that may change from step to step in the proof.

The author would like to thank Oscar Blasco and Petr Honzík for helpful discussions, and Martin Buntinas for the reference [**8**].

2. Lacunary Taylor series

2.1. Fock spaces. A version of the following lemma is found in [**2**] and is needed for the proof of the generalization of Theorem C for F^p_α. We remark that the lemma is similar to Lemma 3 in [**10**], but in this case the domain of integration is over a finite interval instead of $[0, \infty)$.

LEMMA 2.1. *Let $p, \alpha > 0$, and let $\{n_k\}$ be a lacunary sequence. Then for every k,*

$$c\left(\frac{n_k!}{\alpha^{n_k}}\right)^{\frac{p}{2}} n_k^{-\frac{p}{4}+\frac{1}{2}} \leq \int_{\sqrt{\frac{n_k}{\alpha}}}^{\sqrt{\frac{n_{k+1}}{\alpha}}} r^{n_k p} e^{-\frac{\alpha p}{2} r^2}\, r\, dr \leq C\left(\frac{n_k!}{\alpha^{n_k}}\right)^{\frac{p}{2}} n_k^{-\frac{p}{4}+\frac{1}{2}},$$

where c and C are constants independent of k.

PROOF. The second inequality follows directly from Lemma 3 in [**10**]. For the first inequality, we note that since $\{n_k\}$ is lacunary,

$$\frac{n_{k+1}p}{2} \geq \frac{\lambda n_k p}{2} \geq \frac{n_k p}{2} + \sqrt{\frac{n_k p}{2}}$$

when k is sufficiently large. We observe that for $a > 0$ and $x \in \mathbb{R}$, the function $x \mapsto x^a e^{-x}$ is decreasing on $[a, \infty)$; together with the estimate

$$\left(x + \sqrt{x}\right)^x e^{-x-\sqrt{x}} \sim c\, x^x e^{-x}, \qquad x \to \infty$$

and Stirling's formula, we have

$$\begin{aligned}
\int_{\sqrt{\frac{n_k}{\alpha}}}^{\sqrt{\frac{n_{k+1}}{\alpha}}} r^{n_k p} e^{-\frac{\alpha p}{2} r^2}\, r\, dr &= c\left(\frac{2}{\alpha p}\right)^{\frac{n_k p}{2}} \int_{\frac{n_k p}{2}}^{\frac{n_{k+1} p}{2}} u^{\frac{n_k p}{2}} e^{-u}\, du \\
&\geq c\left(\frac{2}{\alpha p}\right)^{\frac{n_k p}{2}} \int_{\frac{n_k p}{2}}^{\frac{n_k p}{2}+\sqrt{\frac{n_k p}{2}}} u^{\frac{n_k p}{2}} e^{-u}\, du \\
&\geq c\left(\frac{2}{\alpha p}\right)^{\frac{n_k p}{2}} \left(\frac{n_k p}{2} + \sqrt{\frac{n_k p}{2}}\right)^{\frac{n_k p}{2}} e^{-\frac{n_k p}{2} - \sqrt{\frac{n_k p}{2}}} \sqrt{\frac{n_k p}{2}} \\
&\geq c\left(\frac{2}{\alpha p}\right)^{\frac{n_k p}{2}} \left(\frac{n_k p}{2}\right)^{\frac{n_k p}{2}} e^{-\frac{n_k p}{2}} \sqrt{\frac{n_k p}{2}} \\
&\geq c\left(\frac{n_k!}{\alpha^{n_k}}\right)^{\frac{p}{2}} n_k^{-\frac{p}{4}+\frac{1}{2}}.
\end{aligned}$$

□

THEOREM 2.2. *Let $1 \leq p \leq 2$, $\{n_k\}$ be a lacunary sequence, and $f(z) = \sum_{n=0}^{\infty} a_n z^n$ be a function in F_α^p. Then*

$$\left(\sum_{k=1}^{\infty} |a_{n_k}|^p \left(\frac{n_k!}{\alpha^{n_k}}\right)^{\frac{p}{2}} n_k^{-\frac{p}{4}+\frac{1}{2}}\right)^{\frac{1}{p}} \leq C\, \|f\|_{F_\alpha^p}$$

for some constant C independent of f.

The finiteness of the sum is already a sufficient condition for membership in F_α^p (see Theorem B, part (i)). We thus obtain a characterization for a function with lacunary Taylor series to belong to F_α^p, $1 \leq p \leq 2$.

PROOF. Let $f \in F_\alpha^p$. By Hölder's inequality and breaking the domain of integration, we have

$$\begin{aligned}
\|f\|_{F_\alpha^p}^p &= c\int_0^{\infty} M_p(r,f)^p\, e^{-\frac{\alpha p}{2} r^2}\, r\, dr \\
&\geq c\int_0^{\infty} M_1(r,f)^p\, e^{-\frac{\alpha p}{2} r^2}\, r\, dr \\
&\geq c\sum_{k=1}^{\infty} \int_{\sqrt{\frac{n_k}{\alpha}}}^{\sqrt{\frac{n_{k+1}}{\alpha}}} M_1(r,f)^p\, e^{-\frac{\alpha p}{2} r^2}\, r\, dr.
\end{aligned}$$

Note that we have

$$M_1(r,f) \geq |a_n|\, r^n$$

for every n, by a calculation involving Cauchy's formula for a_n. Continuing with the calculation, we have

$$\begin{aligned}\|f\|_{F^p_\alpha}^p &\geq c\sum_{k=1}^{\infty}|a_{n_k}|^p\int_{\sqrt{\frac{n_k}{\alpha}}}^{\sqrt{\frac{n_{k+1}}{\alpha}}} r^{pn_k}e^{-\frac{\alpha p}{2}r^2}r\,dr\\ &\geq c\sum_{k=1}^{\infty}|a_{n_k}|^p\left(\frac{n_k!}{\alpha^{n_k}}\right)^{\frac{p}{2}} n_k^{-\frac{p}{4}+\frac{1}{2}};\end{aligned}$$

the last line follows from Lemma 2.1. □

We shall prove a sufficiency condition for membership in F^p_α when $p \geq 2$ using Theorem 2.2 and a duality argument. We first prove the following lemma.

LEMMA 2.3. *Let $p > 0$, $f(z) = \sum_{k=1}^{\infty} a_k z^k$ be an entire function, and $s_n(z) = \sum_{k=1}^{n} a_k z^k$, $n = 1, 2, \ldots$ be its Taylor polynomials. If $\|s_n\|_{F^p_\alpha}$ is bounded above for all n, then $f \in F^p_\alpha$.*

PROOF. Let $p > 0$. On the disk $|z| \leq R$, the function f is continuous and the polynomials s_n converge to f uniformly. Thus $\{s_n\}$ is uniformly bounded on $|z| \leq R$. We can apply the dominated convergence theorem to conclude

$$\begin{aligned}\int_{|z|\leq R}|f(z)|^p e^{-\frac{\alpha p}{2}|z|^2}\,dA(z) &= \lim_{n\to\infty}\int_{|z|\leq R}|s_n(z)|^p e^{-\frac{\alpha p}{2}|z|^2}\,dA(z)\\ &\leq \sup_n\int_{\mathbb{C}}|s_n(z)|^p e^{-\frac{\alpha p}{2}|z|^2}\,dA(z) \leq C.\end{aligned}$$

The result follows by letting $R \to \infty$. □

THEOREM 2.4. *Let $2 \leq p < \infty$, and let $\{n_k\}$ be a lacunary sequence. If $\{a_k\}$ is a sequence of numbers such that*

$$\sum_{k=1}^{\infty}|a_k|^p\left(\frac{n_k!}{\alpha^{n_k}}\right)^{\frac{p}{2}} n_k^{-\frac{p}{4}+\frac{1}{2}} < \infty,$$

then the function $f(z) = \sum_{k=1}^{\infty} a_k z^{n_k}$ is in F^p_α.

PROOF. Let $p' = p/(p-1)$ be the conjugate index of p. For each positive integer N, let $s_N(z) = \sum_{k=1}^{N} a_k z^{n_k}$. Then $s_N \in F^p_\alpha$. Since the dual space of F^p_α is $F^{p'}_\alpha$, up to an equivalence of norms, we have

$$\|s_N\|_{F^p_\alpha} \leq C\sup\left|\int_{\mathbb{C}} s_N(z)\overline{g(z)}e^{-\alpha|z|^2}\,dA(z)\right|$$

for some constant C, where the supremum is taken over all functions g in $F^{p'}_\alpha$ with $\|g\|_{F^{p'}_\alpha} \leq 1$.

Let $g(z) = \sum_{n=0}^{\infty} c_n z^n$ be an entire function in $F_\alpha^{p'}$ with $\|g\|_{F_\alpha^{p'}} \leq 1$. Then by Hölder's inequality and Theorem 2.2,

$$\begin{aligned}
&\left| \int_{\mathbb{C}} s_N(z)\overline{g(z)} e^{-\alpha|z|^2}\, dA(z) \right| \\
&= C \left| \int_0^\infty \sum_{k=1}^N a_k \overline{c_{n_k}} r^{2n_k} e^{-\alpha r^2}\, r\, dr \right| \leq C \sum_{k=1}^N |a_k c_{n_k}| \frac{n_k!}{\alpha^{n_k}} \\
&\leq C \left(\sum_{k=1}^N |a_k|^p \left(\frac{n_k!}{\alpha^{n_k}} \right)^{\frac{p}{2}} n_k^{-\frac{p}{4}+\frac{1}{2}} \right)^{\frac{1}{p}} \left(\sum_{k=1}^N |c_{n_k}|^{p'} \left(\frac{n_k!}{\alpha^{n_k}} \right)^{\frac{p'}{2}} n_k^{-\frac{p'}{4}+\frac{1}{2}} \right)^{\frac{1}{p'}} \\
&\leq C \|g\|_{F_\alpha^{p'}} \left(\sum_{k=1}^\infty |a_k|^p \left(\frac{n_k!}{\alpha^{n_k}} \right)^{\frac{p}{2}} n_k^{-\frac{p}{4}+\frac{1}{2}} \right)^{\frac{1}{p}},
\end{aligned}$$

where C is a constant independent of N. Applying the hypothesis and taking the supremum yield $\|s_N\| \leq C$, and the theorem follows by Lemma 2.3. □

We summarize the results of Theorems B, 2.2 and 2.4 as follows:

THEOREM (Summary). *Let $1 \leq p < \infty$, and let $\{n_k\}$ be a lacunary sequence. A necessary and sufficient condition for the function $f(z) = \sum_{k=1}^\infty a_k z^{n_k}$ to belong to F_α^p is*

$$\sum_{k=1}^\infty |a_k|^p \left(\frac{n_k!}{\alpha^{n_k}} \right)^{\frac{p}{2}} n_k^{-\frac{p}{4}+\frac{1}{2}} < \infty.$$

Furthermore, if $\{n_k\}$ is an arbitrary sequence, then
(i) for $1 \leq p \leq 2$, the sufficiency part holds;
(ii) for $2 \leq p < \infty$, the necessity part holds.

2.2. Bergman spaces. For $0 < p < \infty$, the *Bergman space* A^p consists of those f analytic on the unit disk $\mathbb{D}$ such that

$$\|f\|_{A^p} = \left(\int_0^1 M_p(r,f)^p\, r\, dr \right)^{\frac{1}{p}} = \left(\int_{\mathbb{D}} |f(z)|^p\, dA(z) \right)^{\frac{1}{p}} < \infty,$$

where $dA(z)$ is the Lebesgue area measure. The following theorem concerning functions in A^p with lacunary Taylor series was proved by Buckley, Koskela and Vukotić [**3**].

THEOREM E. *Let $1 \leq p < \infty$, and let $\{n_k\}$ be a lacunary sequence. A necessary and sufficient condition for the function $f(z) = \sum_{k=1}^\infty a_k z^{n_k}$ to belong to A^p is*

$$\sum_{k=1}^\infty |a_k|^p\, n_k^{-1} < \infty.$$

Using the idea of the proof for Theorem 2.2, we give a new proof below of Theorem E for the case $1 \leq p \leq 2$. As in the case of F_α^p, the Hardy–Littlewood theorem for A^p (see [**6**], Chapter 3) already gives the sufficiency part of Theorem E for $1 \leq p \leq 2$ and the necessity part for $2 \leq p < \infty$.

THEOREM 2.5. *Let* $1 \le p \le 2$, $\{n_k\}$ *be a lacunary sequence, and* $f(z) = \sum_{n=0}^{\infty} a_n z^n$ *be a function in* A^p. *Then*

$$\left(\sum_{k=1}^{\infty} |a_{n_k}|^p n_k^{-1}\right)^{\frac{1}{p}} < C \|f\|_{A^p}$$

for some constant C *independent of* f.

PROOF. Let $f \in A^p$. By Cauchy's formula and breaking the domain of integration, we have

$$\begin{aligned}
\|f\|_{A^p}^p &= c \int_0^1 M_p(r,f)^p \, r \, dr \\
&\ge c \int_0^1 M_1(r,f)^p \, r \, dr \\
&\ge c \sum_{k=1}^{\infty} \int_{1-1/n_k}^{1-1/n_{k+1}} M_1(r,f)^p \, r \, dr \\
&\ge c \sum_{k=1}^{\infty} |a_{n_k}|^p \int_{1-1/n_k}^{1-1/n_{k+1}} r^{pn_k+1} \, dr \\
&= c \sum_{k=1}^{\infty} |a_{n_k}|^p \frac{1}{pn_k + 2} \left(\left(1 - \frac{1}{n_{k+1}}\right)^{pn_k+2} - \left(1 - \frac{1}{n_k}\right)^{pn_k+2} \right).
\end{aligned}$$

The lacunarity of $\{n_k\}$ gives us

$$\begin{aligned}
&\ge c \sum_{k=1}^{\infty} |a_{n_k}|^p \frac{1}{n_k} \left(\left(1 - \frac{1}{\lambda n_k}\right)^{pn_k} - \left(1 - \frac{1}{n_k}\right)^{pn_k} \right) \\
&\ge c \sum_{k=1}^{\infty} |a_{n_k}|^p n_k^{-1} \left(e^{-1/\lambda} - e^{-1}\right) = c \sum_{k=1}^{\infty} |a_{n_k}|^p n_k^{-1}.
\end{aligned}$$

□

We omit the proof for the following sufficiency condition for membership in A^p when $p \ge 2$, which is based on a duality argument similar to that found in Theorem 2.4.

THEOREM 2.6. *Let* $2 \le p < \infty$, *and* $\{n_k\}$ *be a lacunary sequence. If* $\{a_k\}$ *is a sequence of numbers such that*

$$\sum_{k=1}^{\infty} |a_k|^p n_k^{-1} < \infty,$$

then the function $f(z) = \sum a_k z^{n_k}$ *is in* A^p.

3. Mixed norm sequence spaces

We prove the following proposition, which can be viewed as an analogue of the Hausdorff–Young theorem with the domains restricted to lacunary blocks.

PROPOSITION 3.1. *Let $\{n_k\}$ be lacunary, and let $I_k = [n_k, n_{k+1}) \cap \mathbb{N}$ and $A_k = \left\{z \colon \sqrt{\frac{n_k}{\alpha}} \leq |z| < \sqrt{\frac{n_{k+1}}{\alpha}}\right\}$. Then for any entire function f,*

$$\sup_{n \in I_k} |a_n| \sqrt{\frac{n!}{\alpha^n}}\, n^{\frac{1}{4}} \leq C \int_{A_k} \left|f(re^{i\theta})\right| e^{-\frac{\alpha}{2}|z|^2}\, dA(z)$$

for some constant C independent of k.

PROOF. Let $f(z) = \sum_{n=0}^{\infty} a_n z^n$. We mimic the proof of the Hausdorff–Young Theorem for Fock spaces (see [**10**]) and look at the integral

$$\int_{A_k} \overline{z}^n f(z) e^{-\alpha|z|^2}\, dA(z).$$

For all sufficiently large k and $n_k \leq n < n_{k+1}$,

$$\begin{aligned}\left|\int_{A_k} \overline{z}^n f(z) e^{-\alpha|z|^2}\, dA(z)\right| &\leq \sup_{z \in A_k} |z|^n e^{-\frac{\alpha}{2}|z|^2} \int_{A_k} |f(z)|\, e^{-\frac{\alpha}{2}|z|^2}\, dA(z) \\ &= \left(\frac{n}{\alpha e}\right)^{\frac{n}{2}} \int_{A_k} |f(z)|\, e^{-\frac{\alpha}{2}|z|^2}\, dA(z).\end{aligned}$$

For the other direction, we calculate

$$\begin{aligned}\left|\int_{A_k} \overline{z}^n f(z) e^{-\alpha|z|^2}\, dA(z)\right| &= c\, |a_n| \int_{\sqrt{n_k/\alpha}}^{\sqrt{n_{k+1}/\alpha}} r^{2n} e^{-\alpha|z|^2}\, r\, dr \\ &= c\, |a_n| \frac{1}{\alpha^n} \int_{n_k}^{n_{k+1}} u^n e^{-u}\, du.\end{aligned}$$

Now fix an integer k, and let $J_k = [n_k, n_{k+1})$, and for each $n \in I_k$, let $\tilde{J}_k$ be the following interval:

$$\tilde{J}_k = \left[\max\left\{n_k, n - \sqrt{n}\right\}, \min\left\{n_{k+1}, n + \sqrt{n}\right\}\right).$$

Then $\tilde{J}_k \subset J_k$, and since $\{n_k\}$ is lacunary, the width of the intervals $\tilde{J}_k$ is comparable to a constant multiple of $\sqrt{n}$ when k is sufficiently large. The real-valued function $\phi(u) = u^n e^{-u}$ has maximum at $u = n$ and points of inflection at $u = n \pm \sqrt{n}$, so that for every $n \in I_k$, the function ϕ is concave down on $\tilde{J}_k$. Thus the area under the graph of ϕ on J_k can be estimated from below by the area of a triangle with base $\tilde{J}_k$ and height $\phi(n)$; that is,

$$\int_{n_k}^{n_{k+1}} u^n e^{-u}\, du \geq c \left(\frac{n}{e}\right)^n \sqrt{n}, \qquad n \in I_k,$$

for some constant c independent of k and n. Thus

$$\left|\int_{A_k} \overline{z}^n f(z) e^{-\alpha|z|^2}\, dA(z)\right| \geq c\, |a_n| \left(\frac{n}{\alpha e}\right)^n \sqrt{n}.$$

Combining the two inequalities we obtain, for all large k and $n \in I_k$,

$$|a_n| \left(\frac{n}{\alpha e}\right)^{\frac{n}{2}} n^{\frac{1}{2}} \leq C \int_{A_k} |f(z)|\, e^{-\frac{\alpha}{2}|z|^2}\, dA(z),$$

and the result follows from the fact that as $n \to \infty$,

$$\left(\frac{n}{\alpha e}\right)^{\frac{n}{2}} n^{\frac{1}{2}} \sim c \sqrt{\frac{n!}{\alpha^n}}\, n^{\frac{1}{4}}.$$

□

COROLLARY 3.2. *Let $\alpha > 0$, $\{n_k\}$ be a lacunary sequence, and $I_k = [n_k, n_{k+1}) \cap \mathbb{N}$. For every $f(z) = \sum_{n=0}^{\infty} a_n z^n \in F_\alpha^1$, we have*

$$\sum_{k=1}^{\infty} \sup_{n \in I_k} |a_n| \sqrt{\frac{n!}{\alpha^n}}\, n^{\frac{1}{4}} \leq C \|f\|_{F_\alpha^1}$$

for some constant C; that is, a necessary condition for an entire function f to belong to F_α^1 is that its coefficients $\{a_n\}$ satisfy

$$\left\{ a_n \sqrt{\frac{n!}{\alpha^n}}\, n^{\frac{1}{4}} \right\} \in \ell^{\infty,1}.$$

PROOF. This follows immediately from the Proposition by summing over k. □

We remark that since $\ell^{\infty,1} \subset \ell^\infty$, this is an improvement of the know Hausdorff–Young inequality (see Theorem A, part (i)). We also remark that the following implication holds

$$f \in F_\alpha^1 \Longrightarrow a_n = o\left(\sqrt{\frac{\alpha^n}{n!}}\, n^{-\frac{1}{4}} \right),$$

which improves the big-O condition in Corollary 5 of [**10**].

We use interpolation theory to fill in the result between F_α^1 and F_α^2. In the following proof we will use the *weighted* spaces $\ell^{p,q}(\mu)$, where where μ is the discrete measure

$$\mu(\{0\}) = 1, \qquad \mu(\{n\}) = \frac{1}{\sqrt{n}}, \qquad n = 1, 2, \ldots.$$

The spaces $\ell^{p,q}(\mu)$ can be thought of as an L^q space consisting of $\ell^p(\mu)$-valued functions, and a sequence $\{a_n\}$ is in $\ell^{p,q}(\mu)$ if

$$\|\{a_n\}\|_{\ell^{p,q}(\mu)} = \left(\sum_{k=1}^{\infty} \left(\sum_{n \in I_k} |a_n|^p \frac{1}{\sqrt{n}} \right)^{\frac{q}{p}} \right)^{\frac{1}{q}} < \infty;$$

The space $\ell^{\infty,1}(\mu)$ coincides with $\ell^{\infty,1}$ because μ is a discrete measure, and the essential sup norm defined by μ is the same as that defined by the counting measure.

THEOREM 3.3. *Let $1 < p \leq 2$, and p' be its conjugate index. Let $\{n_k\}$ be a lacunary sequence, and $I_k = [n_k, n_{k+1}) \cap \mathbb{N}$. For every $f(z) = \sum_{n=0}^{\infty} a_n z^n \in F_\alpha^p$, we have*

$$\left(\sum_{k=1}^{\infty} \left(\sum_{n \in I_k} |a_n|^{p'} \left(\frac{n!}{\alpha^n} \right)^{\frac{p'}{2}} n^{\frac{p'}{4} - \frac{1}{2}} \right)^{\frac{p}{p'}} \right)^{\frac{1}{p}} \leq C \|f\|_{F_\alpha^p};$$

that is,

$$\left\{ a_n \sqrt{\frac{n!}{\alpha^n}}\, n^{\frac{1}{4}} \right\} \in \ell^{p',p}(\mu).$$

We remark that this is an improvement of the Hausdorff–Young theorem (Theorem A, part (i)), since $\ell^{p',p}(\mu) \subset \ell^{p'}(\mu)$.

PROOF. For $f(z) = \sum_{n=0}^{\infty} a_n z^n$ in F^p_α, let T be the operator

$$Tf = \left\{ a_n \sqrt{\frac{n!}{\alpha^n}}\, n^{\frac{1}{4}} \right\}.$$

The well-known equation

$$\|f\|^2_{F^2_\alpha} = \sum_{n=0}^{\infty} |a_n|^2 \frac{n!}{\alpha^n},$$

shows T is a bounded operator from F^2_α to $\ell^2(\mu) = \ell^{2,2}(\mu)$. Corollary 3.2 shows T is a bounded operator from F^1_α to $\ell^{\infty,1}(\mu)$.

We now use the complex interpolation method (see Chapter 4 of [**4**]) to get the desired result. More specifically, Janson, Peetre and Rochberg ([**7**], Theorem 7.3) give the interpolation spaces between the various F^p_α spaces: for $0 < \theta < 1$,

$$[F^2_\alpha, F^1_\alpha]_\theta = F^p_\alpha,$$

where $\frac{1}{p} = (1-\theta)/2 + \theta$. For the interpolation spaces between $\ell^{2,2}(\mu)$ and $\ell^{\infty,1}(\mu)$, we apply the theorem concerning interpolation between vector-valued L^p spaces (see 5.1.2 in Bergh and Löfström [**4**]) to conclude that

$$[\ell^{2,2}(\mu), \ell^{\infty,1}(\mu)]_\theta = \ell^{q,p}(\mu),$$

where $\frac{1}{p} = (1-\theta)/2 + \theta$ and $\frac{1}{q} = (1-\theta)/2 = \frac{1}{p'}$, so that q is simply the conjugate index of p. We apply complex interpolation method to conclude that T is a bounded operator from F^p_α to $\ell^{p',p}(\mu)$ for $1 \le p \le 2$. □

The following sufficient condition for membership in F^p_α when $p \ge 2$ can be proved by a duality argument, and the proof is omitted.

THEOREM 3.4. *Let $2 \le p < \infty$, and p' be its conjugate index. Let $\{n_k\}$ be a lacunary sequence, and $I_k = [n_k, n_{k+1}) \cap \mathbb{N}$. For every sequence $\{a_n\}$ such that*

$$\left\{ a_n \sqrt{\frac{n!}{\alpha^n}}\, n^{\frac{1}{4}} \right\} \in \ell^{p',p}(\mu),$$

we have $f(z) = \sum_{n=0}^{\infty} a_n z^n \in F^p_\alpha$, with

$$\|f\|_{F^p_\alpha} \le C \left(\sum_{k=1}^{\infty} \left(\sum_{n \in I_k} |a_n|^{p'} \left(\frac{n!}{\alpha^n}\right)^{\frac{p'}{2}} n^{\frac{p'}{4}-\frac{1}{2}} \right)^{\frac{p}{p'}} \right)^{\frac{1}{p}}.$$

4. Coefficient Multipliers

THEOREM 4.1. *A sufficient condition for a sequence of complex numbers $\{\lambda_n\}$ to be a multiplier from F^1_α to ℓ^1 is $\left\{\lambda_n \sqrt{\frac{\alpha^n}{n!}}\, n^{-\frac{1}{4}}\right\} \in \ell^{1,\infty}$; that is,*

$$\sup_k \sum_{n \in I_k} |\lambda_n| \sqrt{\frac{\alpha^n}{n!}}\, n^{-\frac{1}{4}} < \infty. \tag{2}$$

The exponent $-\frac{1}{4}$ is sharp.

PROOF. Let $f(z) = \sum_{n=0}^{\infty} a_n z^n$ be in F_α^1. Then

$$\begin{aligned}\sum_{n=I_0}^{\infty} |\lambda_n a_n| &= \sum_{k=1}^{\infty}\left(\sum_{n\in I_k} |\lambda_n| \sqrt{\frac{\alpha^n}{n!}}\, n^{-\frac{1}{4}}\, |a_n| \sqrt{\frac{n!}{\alpha^n}}\, n^{\frac{1}{4}}\right)\\ &\leq \sum_{k=1}^{\infty}\left(\sup_{n\in I_k} |a_n| \sqrt{\frac{n!}{\alpha^n}}\, n^{\frac{1}{4}}\right)\left(\sum_{n\in I_k} |\lambda_n| \sqrt{\frac{\alpha^n}{n!}}\, n^{-\frac{1}{4}}\right)\\ &\leq C\sum_{k=1}^{\infty}\left(\int_{A_k} |f(re^{i\theta})|\, e^{-\frac{\alpha}{2}|z|^2}\, dA(z)\right)\left(\sum_{n\in I_k} |\lambda_n| \sqrt{\frac{\alpha^n}{n!}}\, n^{-\frac{1}{4}}\right),\end{aligned}$$

by Proposition 3.1. Continuing with the estimate and applying the hypothesis, we have

$$\begin{aligned}\sum_{n=I_0}^{\infty} |\lambda_n a_n| &\leq C\left(\sup_k \sum_{n\in I_k} |\lambda_n| \sqrt{\frac{\alpha^n}{n!}}\, n^{-\frac{1}{4}}\right)\left(\sum_{k=1}^{\infty}\int_{A_k} |f(re^{i\theta})|\, e^{-\frac{\alpha}{2}|z|^2}\, dA(z)\right)\\ &\leq C\, \|f\|_{F_\alpha^1} < \infty.\end{aligned}$$

We now show that the exponent $-\frac{1}{4}$ is sharp; that is, given any $\varepsilon > 0$, there is a sequence $\{\lambda_n\}$ that satisfies

$$\sup_k \sum_{n\in I_k} |\lambda_n| \sqrt{\frac{\alpha^n}{n!}}\, n^{-\frac{1}{4}-\varepsilon} < \infty$$

but is not a multiplier. Let $\varepsilon > 0$, and let $\{\lambda_n\}$ be the lacunary sequence where

$$\lambda_n = \begin{cases} \sqrt{\frac{n!}{\alpha^n}}\, n^{\frac{1}{4}+\varepsilon}, & n = n_k,\\ 0 & \text{otherwise.}\end{cases}$$

Look at the function $f(z) = \sum_{n=0}^{\infty} a_n z^n$, where

$$a_n = \begin{cases} \sqrt{\frac{\alpha^n}{n!}}\, n^{-\frac{1}{4}-\varepsilon/2}, & n = n_k,\\ 0 & \text{otherwise.}\end{cases}$$

Then $f(z)$ is in F_α^1 by Theorem 2.2, but $\sum_n |\lambda_n a_n| = \sum_k (n_k)^{\varepsilon/2} = \infty$. □

REMARK 4.2. The case $p = 1$ of Theorem 2.2 follows as a corollary of the above theorem by letting

$$\lambda_n = \begin{cases} \sqrt{\frac{n!}{\alpha^n}}\, n^{\frac{1}{4}} & n = n_k \text{ for some } k;\\ 0 & \text{otherwise.}\end{cases}$$

In the converse direction, we are able to give a necessary condition that is off by a factor of $\sqrt{n}$. We first prove the following lemma that gives a condition on a sequence to belong to the mixed norm sequence space $\ell^{1,\infty}$. This lemma shows that the results in **[5]** and **[6]** can be described in terms of mixed norm spaces. For the following lemma and theorem, we assume that the mixed-norm spaces are formed by a lacunary sequence $\{n_k\}$ satisfying the addition property that $\{n_{k+1}/n_k\}$ is bounded above.

LEMMA 4.3. *Let c_n be a sequence of nonnegative numbers, and $\delta > 0$. Suppose $\ell^{1,\infty}$ is a mixed-norm space formed from a lacunary sequence $\{n_k\}$ where $\{n_{k+1}/n_k\}$ is bounded above. If $\sum_{n=1}^{N} n^\delta c_n = O(N^\delta)$, then $\{c_n\} \in \ell^{1,\infty}$.*

PROOF. Let n be in the k-th lacunary block, that is, $n_k \leq n < n_{k+1}$. Then

$$Cn_{k+1}^{\delta} \geq \sum_{n=1}^{n_{k+1}} n^{\delta} c_n \geq (n_k)^{\delta} \sum_{n \in I_k} c_n.$$

Since $n_{k+1}/n_k \leq C$, we have $\sum_{n \in I_k} c_n \leq C^{\delta}$, so $\{c_n\} \in \ell^{1,\infty}$. □

THEOREM 4.4. *Let $\alpha > 0$. Suppose $\ell^{1,\infty}$ is a mixed-norm space formed from a lacunary sequence $\{n_k\}$ where $\{n_{k+1}/n_k\}$ is bounded above.*

A necessary condition for a sequence of complex numbers $\{\lambda_n\}$ to be a multiplier from F_α^1 to ℓ^1 is $\left\{\lambda_n \sqrt{\frac{\alpha^n}{n!}}\, n^{-\frac{3}{4}}\right\} \in \ell^{1,\infty}$; that is,

$$\sup_k \sum_{n \in I_k} |\lambda_n| \sqrt{\frac{\alpha^n}{n!}}\, n^{-\frac{3}{4}} < \infty. \tag{3}$$

PROOF. Fix $r < 1$, and let

$$f_r(z) = e^{\frac{\alpha}{2}(rz)^2}, \qquad g_r(z) = z\, e^{\frac{\alpha}{2}(rz)^2}.$$

f_r and g_r are in F_α^1, because they are of growth less than $(2, \alpha/2)$ (see p.15, [**9**]). A calculation shows

$$\|f_r(z)\|_{F_\alpha^1} \leq \frac{C}{(1-r)^{1/2}},$$

and

$$\|g_r(z)\|_{F_\alpha^1} \leq \frac{C}{1-r}.$$

Note that $f_r(z) = \sum_{m=0}^{\infty} \frac{\alpha^m r^{2m}}{2^m m!} z^{2m}$, and $g_r(z) = \sum_{m=0}^{\infty} \frac{\alpha^m r^{2m}}{2^m m!} z^{2m+1}$. Then since $\{\lambda_n\}$ is a multiplier, we have

$$\sum_{m=0}^{\infty} |\lambda_{2m}| \frac{\alpha^m r^{2m}}{2^m m!} \leq C \|f_r\|_{F_\alpha^1} \leq \frac{C}{(1-r)^{1/2}},$$

$$\sum_{m=0}^{\infty} |\lambda_{2m+1}| \frac{\alpha^m r^{2m}}{2^m m!} \leq C \|g_r\|_{F_\alpha^1} \leq \frac{C}{(1-r)}.$$

Now let N be an integer, and choose $r = 1 - 1/N$. Then

$$\sqrt{N} \geq \sum_{m \leq N} |\lambda_{2m}| \frac{\alpha^m (1-1/N)^{2m}}{2^m m!} \geq c \sum_{m \leq N} |\lambda_{2m}| \frac{\alpha^m}{2^m m!}$$

and similarly

$$N \geq c \sum_{m \leq N} |\lambda_{2m+1}| \frac{\alpha^m}{2^m m!}.$$

Then by Lemma 4.3,

$$\left\{\lambda_{2m} \frac{\alpha^m}{2^m m!} m^{-\frac{1}{2}}\right\} \in \ell^{1,\infty}, \qquad \left\{\lambda_{2m+1} \frac{\alpha^m}{2^m m!} m^{-1}\right\} \in \ell^{1,\infty}.$$

Now consider the expression $\sqrt{\frac{\alpha^n}{n!}}\, n^{-\frac{3}{4}}$, as in the statement of the theorem. For large $n = 2m$, the expression is comparable to

$$\sqrt{\frac{\alpha^{2m}}{(2m)!}}\,(2m)^{-\frac{3}{4}} \sim \left(\frac{\alpha e}{2m}\right)^m m^{-1} \sim \frac{\alpha^m}{2^m m!} m^{-\frac{1}{2}}.$$

For large $n = 2m+1$, we have

$$\sqrt{\frac{\alpha^{2m+1}}{(2m+1)!}}\,(2m+1)^{-\frac{3}{4}} \sim \left(\frac{\alpha e}{2m}\right)^m m^{-\frac{3}{2}} \sim \frac{\alpha^m}{2^m m!}\, m^{-1}.$$

Thus

$$\left\{\lambda_{2m}\sqrt{\frac{\alpha^{2m}}{(2m)!}}\,(2m)^{-\frac{3}{4}}\right\} \in \ell^{1,\infty}, \qquad \left\{\lambda_{2m+1}\sqrt{\frac{\alpha^{2m+1}}{(2m+1)!}}\,(2m+1)^{-\frac{3}{4}}\right\} \in \ell^{1,\infty},$$

and the theorem follows by adding the two sequences. □

References

[1] O. Blasco, *Introduction to vector valued Bergman spaces*, in *Function spaces of operator theory (Joensuu, 2003)*, Univ. Joensuu Dept. Math. Rep. Ser. **8** (2005), 9–30.

[2] O. Blasco and A. Galbis, *On Taylor coefficients of entire functions integrable against exponential weights*, Math. Nachr. **223** (2001), 5–21.

[3] S. Buckley, P. Koskela and D. Vukotić, *Fractional integration, differentiation, and weighted Bergman spaces* , Math. Proc. Camb. Phil. Soc. **126** (1999), 369–385.

[4] J. Bergh and J. Löfström, *Interpolation Spaces* , Springer-Verlag, Berlin, 1976.

[5] P. L. Duren, *Theory of H^p Spaces*, Dover Publications, Mineola, New York, 2000.

[6] P. Duren and A. Schuster, *Bergman Spaces*, American Mathematical Society, Providence, RI, 2004.

[7] S. Janson, J. Peetre, and R. Rochberg, *Hankel forms and the Fock space*, Revista Mat. Iberoamericana, **3** (1987), 61–129.

[8] C. N. Kellogg, *An extension of the Hausdorff–Young Theorem*, Michigan Math. J., **18** (1971), 121–127.

[9] J. Tung, *Fock Spaces*, Ph.D. thesis, University of Michigan, 2005.

[10] J. Tung, *Taylor coefficients of functions in Fock spaces*, J. Math. Anal. Appl., **318** (2006), no. 2, 397–409.

Department of Mathematical Sciences, DePaul University, Chicago, Illinois 60614
E-mail address: ytung@depaul.edu

Titles in This Series

TITLES IN THIS SERIES

430 **Idris Assani, Editor,** Ergodic theory and related fields, 2007

429 **Gui-Qiang Chen, Elton Hsu, and Mark Pinsky, Editors,** Stochastic analysis and partial differential equations, 2007

428 **Estela A. Gavosto, Marianne K. Korten, Charles N. Moore, and Rodolfo H. Torres, Editors,** Harmonic analysis, partial differential equations, and related topics, 2007

427 **Anastasios Mallios and Marina Haralampidou, Editors,** Topological algebras and applications, 2007

426 **Fabio Ancona, Irena Lasiecka, Walter Littman, and Roberto Triggiani, Editors,** Control methods in PDE-dynamical systems, 2007

425 **Su Gao, Steve Jackson, and Yi Zhang, Editors,** Advances in Logic, 2007

424 **V. I. Burenko, T. Iwaniec, and S. K. Vodopyanov, Editors,** Analysis and geometry in their interaction, 2007

423 **Christos A. Athanasiadis, Victor V. Batyrev, Dimitrios I. Dais, Martin Henk, and Francisco Santos, Editors,** Algebraic and geometric combinatorics, 2007

422 **JongHae Keum and Shigeyuki Kondō, Editors,** Algebraic geometry, 2007

421 **Benjamin Fine, Anthony M. Gaglione, and Dennis Spellman, Editors,** Combinatorial group theory, discrete groups, and number theory, 2007

420 **William Chin, James Osterburg, and Declan Quinn, Editors,** Groups, rings and algebras, 2006

419 **Dinh V. Huynh, S. K. Jain, and S. R. López-Permouth, Editors,** Algebra and Its applications, 2006

418 **Lothar Gerritzen, Dorian Goldfeld, Martin Kreuzer, Gerhard Rosenberger, and Vladimir Shpilrain, Editors,** Algebraic methods in cryptography, 2006

417 **Vadim B. Kuznetsov and Siddhartha Sahi, Editors,** Jack, Hall-Littlewood and Macdonald polynomials, 2006

416 **Toshitake Kohno and Masanori Morishita, Editors,** Primes and knots, 2006

415 **Gregory Berkolaiko, Robert Carlson, Stephen A. Fulling, and Peter Kuchment, Editors,** Quantum graphs and their applications, 2006

414 **Deguang Han, Palle E. T. Jorgensen, and David Royal Larson, Editors,** Operator theory, operator algebras, and applications, 2006

413 **Georgia M. Benkart, Jens C. Jantzen, Zongzhu Lin, Daniel K. Nakano, and Brian J. Parshall, Editors,** Representations of algebraic groups, quantum groups and Lie algebras, 2006

412 **Nikolai Chernov, Yulia Karpeshina, Ian W. Knowles, Roger T. Lewis, and Rudi Weikard, Editors,** Recent advances in differential equations and mathematical physics, 2006

411 **J. Marshall Ash and Roger L. Jones, Editors,** Harmonic analysis: Calderón-Zygmund and beyond, 2006

410 **Abba Gumel, Carlos Castillo-Chavez, Ronald E. Mickens, and Dominic P. Clemence, Editors,** Mathematical studies on human disease dynamics: Emerging paradigms and challenges, 2006

409 **Juan Luis Vázquez, Xavier Cabré, and José Antonio Carrillo, Editors,** Recent trends in partial differential equations, 2006

408 **Habib Ammari and Hyeonbae Kang, Editors,** Inverse problems, multi-scale analysis and effective medium theory, 2006

For a complete list of titles in this series, visit the
AMS Bookstore at **www.ams.org/bookstore/**.